The Technology-Energy-Environment-Health (TEEH) Chain in China

ALLIANCE FOR GLOBAL SUSTAINABILITY BOOKSERIES
SCIENCE AND TECHNOLOGY: TOOLS FOR SUSTAINABLE DEVELOPMENT

VOLUME 8

Aims and Scope of the Series

The aim of this series is to provide timely accounts by authoritative scholars of the results of cutting edge research into emerging barriers to sustainable development, and methodologies and tools to help governments, industry, and civil society overcome them. The work presented in the series will draw mainly on results of the research being carried out in the Alliance for Global Sustainability (AGS).
The level of presentation is for graduate students in natural, social and engineering sciences as well as policy and decision-makers around the world in government, industry and civil society.

The Technology-Energy-Environment-Health (TEEH) Chain in China

A Case Study of Cokemaking

Edited by

Karen R. Polenske
Massachusetts Institute of Technology,
Cambridge, Massachusetts, U.S.A.

Springer

A C.I.P. Catalogue record for this book is available from the Library of Congress.

ISBN-13 978-90-481-6864-4
ISBN-10 1-4020-4236-1 (e-book)
ISBN-13 978-1-4020-4236-2 (e-book)

Published by Springer,
P.O. Box 17, 3300 AA Dordrecht, The Netherlands.

www.springeronline.com

Printed on acid-free paper

ALLIANCE FOR GLOBAL SUSTAINABILITY

An International Partnership

Alliance for Global Sustainability
International Advisory Board (IAB)

Chairman:

Prof. Francis Waldvogel, President, ETH Board, Switzerland

AGS University Presidents:

Prof. Olaf Kübler, President, Swiss Federal Institute of Technology, Zürich

Prof. Takeshi Sasaki, President, University of Tokyo

Prof. Jan-Eric Sundgren, President, Chalmers University of Technology

Dr. Charles M. Vest, President, Massachusetts Institute of Technology

Members:

Dr. Thomas Connelly, Chief Science and Technology Officer, DuPont

The Hon. President José-María Figueres Olsen, Managing Director, World Economic Forum

Mr. Kazuo Ogura, President, The Japan Foundation

Dr. Hiroyuki Fujimura, Chairman of the Board, Ebara Corporation

Dr. Jay Ihlenfeld, Vice President, Research & Development, 3M

Mr. Lars G. Josefsson, President and Chief Executive Officer, Vattenfall AB

Mr. Lars Kann-Rasmussen, Director, VKR Holding A/S

Dr. Jill Ker Conway, Chairman, Lend Lease Corporation (Retired)

Mr. Masatake Matsuda, Chairman, East Japan Railway Company

Mr. Nobuya Minami, Advisor, Tokyo Electric Power Company, Inc.

Prof. Jakob Nüesch, Honorary Member, International Committee of the Red Cross

Mr. Dan Sten Olsson, CEO, Stena AB

Mr. Motoyuki Ono, Director General, The Japan Society for the Promotion of Science

Mr. Alexander Schärer, President of the Board, SIM LSchäer Stine AG

Dr. Stephan Schmidheiny, President, Avina Foundation

Mr. Norio Wada, President, Nippon Telegraph and Telephone Corporation (NTT)

Ms. Margot Wallström, Member of the European Commission

Prof. Hiroyuki Yoshikawa, President, National Institute of Advanced Industrial Science and Technology

Dr. Martin Zimmerman, Group Vice President, Corporate Affairs, Ford Motor Company (Retired)

Dr. Hans-Rudolf Zulliger, President Stiftung Drittes Millenium, Board of Directors, Amazys Ltd.

February 2004

CONTENTS

PREFACE

This book deals with the important and timely problem of changes in coke production technology and how these changes are affecting energy intensities, the environment, and human health in coal-rich Shanxi Province, China. There are several factors that lend significance to this topic: China is a rising industrial power at the threshold of a market economy, with rapidly increasing domestic demand for energy and steel. The country has large reserves of metallurgical coal, which is used to make metallurgical coke, a key ingredient for iron and steel production and an important earner of foreign exchange for China. In fact, China is the largest producer of metallurgical coke in the world, and Shanxi Province accounts for 40% of this production. But despite its importance to the economy, coke production has drawbacks: without adequate pollution control, coke production is a serious source of environmental degradation and health hazards.

Professor Karen R. Polenske has assembled and led an international, multidisciplinary team of chemical engineers, scientists, economists, and planners to conduct analyses of this multifaceted problem. By developing methodology and diagnostic tools for coupling energy and economic issues with those of pollution and health hazards, this team has laid the foundation for conducting health-impact assessments in other industrial sectors.

Coke, first produced as a byproduct of illuminating gas manufacture, has acquired great importance in iron and steel making. After the depletion of Britain's forests that provided charcoal for the industrial revolution, successful experiments showed that coke could replace charcoal in the blast furnace to separate iron from iron ore. Like charcoal, metallurgical coke is a source of heat, can act as a reductant, has high permeability so that reactions can proceed rapidly, and possesses the physical strength to withstand crushing and attrition in the blast furnace. This book traces the development of cokemaking technology in China from the inefficient and highly polluting indigenous beehive ovens to modern slot ovens that capture chemical byproducts, thereby reducing pollutant emissions.

The research team's ground breaking work began with extensive surveys given to cokemaking plants, designed to compile data on trends in cokemaking technologies, production scale, energy use, and air-pollutant emissions. The eight sample surveys, conducted between 1998 and 2003, included the training of local officials who would be directly in charge of data collection. Information from these surveys on energy use and emissions has formed the basis for evaluating the Technology-Energy-Environment-Health (TEEH) chain proposed by Polenske in 1998. The TEEH framework shows us how changes in technology impact energy use and pollution, which, in turn, affect human health.

The discussion on human exposure to ultrafine particles in the cokemaking sector includes results from the application of optical and laser diagnostic equipment that measures mutagenic and carcinogenic compounds near coke ovens and coke workers' homes. The time-resolved measurements are fast enough to allow many successive measurements within a short period of time, thereby allowing researchers to follow changes in pollutant concentrations during specific events, such as loading and unloading of ovens or the lighting of a cigarette. It is the first time we have seen researchers capable of providing a quantitative background to discussions on the health impacts of different cokemaking technologies.

The authors' socioeconomic analyses show the contradiction that low productivity township and village-owned enterprise (TVE) coke ovens are more profitable than the larger, more technologically advanced state-owned plants. The locally based TVEs provide employment to a large number of low-paid and less-educated rural workers, but because they make little, or no, investment in environmental controls, TVEs can produce profitably so long as there is a continuing strong demand for metallurgical coke.

China is going through an unprecedented industrial and economic expansion. The expansion of the lucrative coke industry has the potential of bringing more advanced, safer, and cleaner technologies to the development of new plants. By highlighting, with their careful analyses and measurements the problems of pollution and health hazards in the environment of Shanxi's coke ovens, the authors provide a valuable service to plant owners and local and national policy makers on urgently needed decisions for improving the environmental control of metallurgical coke plants in China.

János Beér
February 21, 2005

In China, many cokemaking plants have been built in the last twenty years by township and village enterprises (TVE). These plants support the economic growth of the agricultural villages; however, to minimize cokemaking plant construction and operation costs, mostly poor pollution control units have been installed in the TVE cokemaking plants. As a result, people residing near cokemaking plants have suffered serious health damage, and both the agricultural land and the forests around the cokemaking plants have been severely damaged. Thus, the Chinese cokemaking sector must be rebuilt so that it is capable of realizing a good symbiotic relationship between economic development and the environment.

In this book, the authors explain the results of a comprehensive analysis of the Chinese cokemaking sector and the technologies from economic, energy, environment, and human health points-of-view. The analysis has been conducted on the basis of abundant surveys. As a result, the authors of the chapters in the book are able to indicate a possible path to the sustainable development of the Chinese cokemaking sector.

Masayoshi Sadakata
March 20, 2005

ACKNOWLEDGMENTS

Our cokemaking team is comprised of interdisciplinary researchers from the fields of chemical engineering, economics, physics, and planning, which we refer to in the text as the Shanxi Cokemaking Team, or simply as the Cokemaking Team. Note that the affiliations given for the authors here and in the chapters refer to where each person was located at the time he/she conducted the research. In the chapters, I also indicate their current work location. Our team was assisted by numerous people since we started this project in 1997, some of whom I am certain to have missed in these acknowledgments. To those whom I missed, I express my most sincere apologies.

I am most grateful for the numerous contributions made by the author(s) of each chapter, each of whom has been part of the Cokemaking Team—some for a year and a few for the entire duration of the current phase of the project. Our senior advisors on this project, **Professor János M. Beér**, Professor Emeritus of Chemical and Fuel Engineering, Department of Chemical Engineering, Massachusetts Institute of Technology (MIT) and **Professor Masayoshi Sadakata**, Department of Chemical Systems Engineering, the University of Tokyo, gave us constant guidance on our project. As of Spring 2005, Professor Sadakata is in the Department of Environmental Chemical Engineering, Kogakuin University. I am personally deeply indebted to János who was always available and who had endless patience in answering my numerous technical questions about combustion of coal, particulate pollution, and other technical issues related to the cokemaking sector. He made the subject exciting and challenging to someone with a degree in economics. If I ever decide to obtain a degree from chemical engineering, I would want to be a student of someone as knowledgeable as he is.

Prior to March 20, 2002, members of the Cokemaking Team had published a number of individual and joint papers on our work. At the 2002 cokemaking workshop at the annual Alliance for Global Sustainability (AGS) meetings in San Jose, Costa Rica, I raised with team members the possibility of writing a book. We agreed to do this, based upon an outline of the chapters I presented to them. Soon thereafter, Kluwer Academic Publishers accepted a prospectus I submitted for consideration as part of their Alliance for Global Sustainability Bookseries: Science and Technology Tools for Sustainable Development.

I am deeply indebted in many ways to **Professor Chen Xikang**, Chinese Academy of Sciences, Academy of Mathematics and Systems Science, Beijing. He was the co-editor with me of my first book on China, entitled, *Chinese Economic Planning and Input-Output* Analysis, published in 1991 by Oxford University Press. Since my first trip to China in 1986, he and I have collaborated on numerous projects, and I had the pleasure of working with him on this project. Without his help in making many detailed arrangements for our field trips, sending information in response to numerous questions my team and I continued to raise concerning statistics, regulations, and details of cokemaking, as well as providing me with wonderful green tea from Heilongjiang Province to keep me young at heart, our cokemaking team would never have had such unique access to information in Beijing and Shanxi Province. During the past six years, **Professor Yang Cuihong** has worked closely with Chen Xikang and other members of the cokemaking team on this project, having completed her PhD in May 1999. She accompanied us on most of our field trips, helped design and implement the surveys, and wrote and presented papers at the various AGS annual meetings and other meetings in order to let others in China know of our work. Given her knowledge and sophistication in doing this research, I regarded her as a colleague rather than a student, even when she was still a PhD student. Both Xikang and Cuihong worked throughout the project on many aspects of our research, but they had primary responsibility for the coke plant and local official surveys. Their details from those surveys, summarized in Chapter 2, benefited almost all members of the Cokemaking Team in their own research. I thank both of them, as well as their colleague, Li Jinghua, for their significant and critical contributions to this book.

Professor Steven B. Kraines, Associate Professor, Department of Frontier Sciences and Science Integration, the University of Tokyo, was another of the original members of the cokemaking team. He came with us on all but one of the field trips, helped extensively in writing the proposals for the funding, and provided excellent support throughout the project. His willingness to work on the transport issues during the initial stages of our project provided a critical set of analyses that members of my team at MIT were able to take over during the last few years, some of which is documented in Chapter 6.

Professor Hans C. Siegmann, Professor Siegmann of Physics at the Swiss Federal Institutes of Technology, Zürich, now at the Stanford Center for the Study of Ultrafast Phenomena, also is an original member of the cokemaking team. He contributed his extensive knowledge of particulate air pollution, and prior to our project, he designed two of the three sensors we used for measuring the particulate air pollution initially at the coke plants, but soon also on the highways (from diesel trucks), and in the coke worker homes (Chapter 7). Especially significant were people with whom he worked on the project, several of whom I reference below. Here, I highlight the continuous contribution made by **Dr. Qian Zhiqiang,** Senior Research Scientist, Laboratory for Solid State Physics, Swiss Federal Institutes of Technology, Zürich, and Taiyuan University of Technology. He either personally did the particulate-pollution monitoring for us or supervised others. His humor helped to lighten each day on our field trips, and his own understanding of the field conditions and physics complemented Hans' extensive knowledge of the physics underlying the sensors we used. Both of them helped me understand the significance of our work in the larger scope of global particulate-pollution monitoring and to converse intelligently with the plant managers and officials from the Shanxi Environmental Pollution Bureau. Although **Dr. Uli Matter** never accompanied us in person on the field trips, his work in Switzerland to test the sensors to ensure that they were ready for use in the field was critical for the overall success of our project.

Professor Fang Jinghua, Department of Thermal Engineering, Taiyuan University of Technology, deserves special mention not only for the help he provided Qian on the pollution monitoring, but equally important for the numerous arrangements he made with key officials, plant managers, and other relevant people for each of our field trips in Shanxi Province. Frequently, he would be able to get us arrangements with plant managers who were busy and who often could not understand why we wanted to visit their plant yet again. He had to explain the nature of our inquiries to them. We were fortunate that he had spent three years at MIT prior to the start of our project and was therefore aware of the specificity we needed in order to produce excellent papers and an excellent book. He assisted with the translations from time-to-time as well.

Although **Hoi-Yan Erica Chan, Chen Hao, Chen Yan, Holly Krambeck, Guo Wei, and Dr. Ali Shivani-Mahdavi** were MIT students at the time they initially worked on the project, all of them put in significant hours of work during the past two years to redo their chapters (3, 4, 5, 6, and 8) so that the information would be up-to-date. Ali and Guo Wei did additional calculations that took many hours. Ali also helped during the last weeks of the writing of this manuscript in reviewing key parts of the book. I especially appreciate the way in which during the past six months Ali constantly asked if there was any additional help he could provide. Although Erica and Holly had not worked at the same time on the project at MIT, they wrote an interesting Chapter 8 based both on the diversity of their experiences on this project and their own considerable expertise and insights into the socioeconomic conditions. Chen Hao did the case studies that helped us compare different coke ovens in detail, and Chapter 3 is based upon research he did for his master's thesis, supplemented by the vast amounts of information I have collected since he completed his thesis in 2000.

In addition to all of these significant inputs by the authors of the chapters, I thank the following who helped me directly, or who assisted indirectly by working with the chapter authors.

CHINA

<u>Chinese Academy of Sciences, Academy of Mathematics and Systems Science, Beijing</u>

One of the primary groups working with us on this project were Academy of Mathematics and Systems Science (AMSS) staff from the Chinese Academy of Sciences (CAS) who assisted Professors Chen Xikang and Yang Cuihong and Dr. Li Jinghua on the surveys and other details we needed for the research, including:

Hu Guirong, Associate Professor, Shanxi University, who conducted field-survey and training programs in 1998, when she was a visiting scholar with the Academy of Mathematics and Systems Science.

Liu Xinjian, Professor, Yanshan University, Hebei Province, who conducted field surveys in 1998 and pollution monitoring in 1999.

Pan Xiaoming, Associate Professor, Academy of Mathematics and Systems Science, who conducted field surveys and wrote survey reports from 1997 to 1999.

Xu Yiping, PhD, who conducted pollution monitoring in 1999.

We also thank the following Academy of Mathematics and Systems Science PhD students, who helped compile survey results:

- **He Jing,**
- **Liu Xiuli,**
- **Wang Eryuan,**
- **Xu Jian,**
- **Xu Xiangmin,** and
- **Zhang Hongxia.**

<u>**Shanxi Province**</u>

We could not have conducted the surveys, made the field trips, and had many of the other interactions with those related to Shanxi Province cokemaking without the extensive cooperation and assistance from numerous Shanxi Province Officials. We especially thank the following:

Li Shaojing, Professor and Senior Engineer, Shanxi Environmental Protection Bureau, who met with us frequently and gave us invaluable information regarding pollution standards and practices in the Shanxi cokemaking industry.

Guan Cunxian, Deputy director of Shanxi EPB, who accompanied us on our field trip as we visited a large number of coke plants for the first time and who helped us organize the cokemaking surveys in Shanxi Province, starting in 1998.

For helping us with the surveys, we especially thank the following additional Shanxi Province officials and staff:

- **Hao Shuiming,**
- **Meng Yulin,**
- **Sun Quan,**
- **Xu Gang,**
- **Zhang Shuanggui**, and
- **Zhao Zhijie.**

Qian Zhiqiang is grateful for the assistance with the air pollution monitoring contributed by Professors Fang Jinghua (Taiyuan Institute of Technology) and Yang Cuihong (CAS), throughout the entire project. He also thanks the following:

Chinese Academy of Sciences:

- **Liu Xingjian** (former PhD student), and
- **Xu Yiping** (former PhD student).

Taiyuan University of Technology:

- **Tan Wenyuan,** Associate Professor,

- **Zhang Pangshen**, Engineer,
- **Wang Ruoxia**, Associate Engineer,

and the following graduate students:

- **Duan Qiong,**
- **Li Hongge,**
- **Li Yabing,**
- **Su Huarong,**
- **Wu Hui,**
- **Zhang Yongsheng**, and
- **Zhen Xianrong.**

In addition, he is grateful for the assistance provided by the coke plant managers and workers in giving him and his measurement team access to the plants and homes in order to conduct the monitoring.

We also thank Shanxi Province technical experts in the field of cokemaking, **Shen Weiqing** and **Wu Xiaoping,** who provided the Cokemaking Team valuable information about cokemaking in the region.

Specific Coal Mines and Coke Plants

Through single and often repeated visits to many coke plants in Shanxi Province, and also a few in other parts of China, as well as in Japan and the United States, we were able to uncover the detailed invaluable information that helps make this book unique. Managers and others at the following plants, mainly located in Shanxi Province, China, unless otherwise indicated, met with members of our team, provided detailed data to us both orally and in written form, and helped to educate us as to the important, lately dramatic, changes occurring in the making of coke:

Dongshan Coal Mine Company
Gangyuan Coking Company, Ltd., Qingxu County
Kimitzu Coke Plant, Nippon Steel Chemical Co., Ltd., Kimitsu, Japan
Meijin Group, Qingxu County
Qingxu Coal Gasification General Company, South Town of Taiyuan, Qingxu County
Sanjia Cokemaking Plant, Mianshan Tourist Site (Jiexiu)
Sansheng Coking Co., Ltd., Jiexiu
Shanxi Antai Group, Jiexiu (http://www.antaigroup.com/)
Shanxi Coking Industrial (Group) Corp., Hongtong
Shanxi Electric Power Company, Taiyuan
Shanxi Houma Pilot Plant, Houma
Shanxi Sanjia Coal Chemistry Co., Ltd., Jiexiu (http://www.cnsjcoke.com/)
Shenango Inc., Neville Island, Pittsburgh, PA, USA,
Sun Coke Company, Indiana Harbor, IN, USA
Taiyuan Coal Gasification (Group) Corporation, Ltd., Taiyuan
Taiyuan Dongsheng Coking Company, Ltd., Qingxu County
Taiyuan Iron and Steel Corporation, Taiyuan
Taiyuan No. 1 Heat and Power Plant, Taiyuan
Taiyuan Yingxian Coal-Carbonization Group Co., Ltd., Qingxu

US Steel Clairton Works, Clairton, PA, USA
Xishan Coal Gasification Company, Guijiao

Specific Agencies, Bureaus, Commissions, Design Institutes, and Universities.
We also thank the many people who generously gave of their time at the following.
Anshan Coking & Refractory Engineering Consulting Corporation (ACRE), Anshan
China Coal & Coke Holding, Ltd., Beijing
China Coking Industry Association, Beijing
Chinese Research Academy of Environmental Sciences, Beijing
Institute of Coal Chemistry, Chinese Academy of Sciences, Taiyuan
Institute of Comprehensive Transportation of State Development Planning Commission, Beijing
Northern Jiaotong University, Beijing
Shanxi Chemical Design Institute, Taiyuan
Shanxi Design Institute of Transport Planning and Survey, Taiyuan
Shanxi Development and Reform Commission, Taiyuan
Shanxi Economic and Trade Commission, Taiyuan
Shanxi Environmental Protection Bureau, Taiyuan
Shanxi Province Chemical Design Institute, Taiyuan
Shanxi Township Enterprises Coke Supply and Marketing, Ltd., Taiyuan
Shanxi Township and Village Bureau, Taiyuan
Shanxi Transportation Bureau, Taiyuan
State Environmental Protection Administration of China (SEPA), Beijing
UNDP/GEF Energy Conservation and Greenhouse Gas Emissions Project Management Office, Beijing

Those who specifically helped us with the surveys and other data collection, included the National TVE Bureau under the Ministry of Agriculture of China, Shanxi Provincial TVE Bureau, and Shanxi Statistical Bureau.

THE UNIVERSITY OF TOKYO

The AGS project was a true collaboration among the participating universities. **Professor Steven Kraines**, the University of Tokyo, headed the University of Tokyo team. During 1999-2001, he was assisted on the work by **Takeyoshi Akatsuka,** who received his master's degree at the University of Tokyo's Department of Chemical Systems Engineering in 2001. Takeyoshi's research for his thesis helped provide some of the initial transportation information that is underpinning Chapter 6.

Other contributors we thank include:

- **Yuji Sakai,** PhD.;
- **Yan Li,** PhD;
- **Jimmy,**
- **Yoshinori Takagi,** and
- **Shigeru Aoki.**

We also thank Professors Shunichi Araki and Ryoji Sakai, who worked with us until the fall of 2000. As noted earlier, throughout the seven-year project, **Professor Masayoshi Sadakata** provided strong encouragement, assisted us with finding funds, and helped us to conduct this important collaborative research.

MASSACHUSETTS INSTITUTE OF TECHNOLOGY

At the Massachusetts Institute of Technology (MIT), the multiregional planning (MRP) research team worked on each part of the present book. I deeply thank our technical assistants **Turi McKinley** and **Dr. Natalia Sizov**, who worked on the photos, PowerPoint presentations, websites, and many other technical details that helped the MIT staff convey our work effectively to others. Based upon the many presentations at MIT and elsewhere in the United States, they helped me improve our presentation material, so that our important work would be appreciated both by technical and lay people. Turi accompanied us on the field trip in 2000, taking videos that our Cokemaking Team used to present the following year to the managers in the coke plants we visited during that trip, and she went with me to the Pittsburgh, PA cokemaking conference in the fall of 2001 to show some of the videos during a three-hour workshop on our cokemaking research that I held at the conference. I gained tremendous insights into our work by viewing the research through the various details she highlighted in her photos and by the comments she provided throughout the field trip. I also thank our numerous research assistants:

- **Ai Ning,**
- **Sharon Chan,**
- **Chen Danwen,**
- **Edward Cheung,**
- **David Greenblatt,**
- **Guo Wei,**
- **Guo Zhan,**
- **Huang Jing,**
- **Holly Krambeck,**
- **Ali Shirvani-Mahdavi,**
- **Li Xin,**
- **Li Yu,**
- **Shi Xiaoyu,**
- **Xu Rongtao,** and
- **Zhang Yabei.**

Of these, I would never have been able to continue with the editing of the book during the past year without the excellent and extensive work by Holly Krambeck, who reviewed each chapter (some several times), rewriting some sections to improve the logic, significantly improving the grammar, checking for places where our meaning was not clear, and pushing me to complete the work. She also prepared the field notes from the trip to Shanxi Province in January of 2004 and did a dry run of the book for the team members at the AGS conference at Chalmers University in March 2004. I thank Xu

Rongtao for his patience in working on the formatting of the various tables and figures in the book, laboriously recreating some that needed to be redone for publication.

Especially during the last few months of preparing the manuscript, I benefited greatly from the work that Li Xin and Shi Xiaoyu did in formatting, reviewing, checking, translating Chinese terms, interfacing with our colleagues in China by phone and fax, and all the other final preparation steps needed to ready the manuscript for publication. Shi Xiaoyu reviewed each chapter, and just when I thought we were finished, she found many places where the meaning was not clear. She also helped me in communicating with the Chinese officials both in Beijing and Shanxi Province, some of whom speak no English. She accompanied me to the 2003 Cokemaking Conference in Beijing and helped to introduce me to many of the coke managers attending the conference. Li Xin spent hours doing the indexing and formatting, checking for consistencies in the headings, arranging all the photos in the appendix, and other important, work required to finish a manuscript. I am deeply indebted to her for being willing to do this for the book and her patience and good humor as she did this task quickly and efficiently. During each of the past four years, Chen Danwen who was in high school volunteered to work with me in the summer. I thank her for scanning the numerous photos after each of my trips, a task that will no longer be necessary with my new digital camera.

Indeed, I have been richly blessed with many talented, intelligent, unusually hard-working research assistants, who seem never to complain when their work with me takes them away from their own weekends with friends.

I also benefited greatly from the various coke conferences at which I have given major speeches, such as the 2000 (Chicago), 2002 (St. Louis), as well as the 2001 (Pittsburgh) and 2003 (Toronto), at which I gave three-hour workshops on our China coke research, each sponsored by Intertech. In addition, I gave one of the main talks at the 2nd China International Coking Technology and Coke Market Congress in Beijing in June 2004, sponsored by the Metallurgical Council, China Council for the Promotion of International Trade (MC - CCPIT). The U.S. conferences were attended by about 200 and the Beijing conference by about 500 coke managers and others working in the coke and related industries from throughout the world. In each case, my talks were given to the entire group attending, and I gained considerable information from my exchanges with these people as well as from their presentations during the two-day conferences.

SWISS FEDERAL INSTITUTES OF TECHNOLOGY, ZÜRICH (ETHZ)

Hans Siegmann worked not only with Uli Matter, and Qian Zhiqiang in Switzerland, but he had other talented associates working with him on the particulate-pollution sensors. We especially thank:

Mr. Leo Scherrer, who worked in the Laboratory for Solid State Physics in Switzerland and built two of the three sensors based on the design by Hans. We used these sensors not only on the various monitoring field trips in Shanxi Province, but Qian also used them for his extensive testing at specific plants. Mr. Scherrer also helped write the instructions for use of the PAH (polycyclic aromatic hydrocarbon) sensor. I personally used those instructions and the PAH sensor extensively during the initial three years of our project.

Dr. Reinhold Wasserkort, MIT, who formerly had worked with Professor Siegmann and Dr. Scherrer in the ETHz laboratory, was fortunately at MIT during the first three years of our project and instructed me in how the PAH sensor operated. He and Hans Siegmann developed in me a great appreciation of the pollution surrounding us in our daily life and the harm especially the ultrafine particulates do to human health.

Last But Not Least

Finally, I thank the following people, who have made valuable contributions to our research:

Professor Claude Lupis, MIT, met with me several times and helped our students in their understanding of the cokemaking technologies and the financing issues connected with bringing new technologies into the sector. I always looked forward to the discussions we would have at our lunches.

Professor Francis C. McMichael, Carnegie Mellon University, worked with us during the year (1999-2000) that he was a visiting professor in the Chemical Engineering department at MIT. He accompanied us on our field trip to Shanxi Province in July 2000, and he worked with several of our research assistants in writing their reports and theses, based upon his extensive knowledge of the details of the U.S. cokemaking technologies in Pittsburgh, Pennsylvania.

Dr. Domenico Maeillo (e-mails with coal, coke, and related information). Each day, Domenico sends me and others e-mails, which helps me keep abreast of the rapid, and currently dramatic, changes occurring in the coal, coke, iron ore, steel, cargo ship, and related sectors throughout the world. Those e-mails have been extremely valuable in keeping me abreast of current events, especially those in the commodity markets.

FUNDING

Each group in our Cokemaking Team (the Chinese, Japanese, Swiss, and U.S. participants) received major funding from the Alliance for Global Sustainability (AGS) during all the seven years of the project. We truly appreciate that sustained financial support. Even so, AGS funding was insufficient for such a large-scale project, especially given our need to do extensive surveys and pollution monitoring and to maintain data bases. We therefore deeply appreciate the financial support from all sources, including:

Support To All Teams:

AGS	Alliance for Global Sustainability

Support to MIT Team:

AGS	AGS Martin Fellowships	small grants
CEEPR	Center for Energy and Environmental Policy	small grants

	Research (MIT)	
CIS	Center for International Studies (MIT)	small grants
DUSP	Department of Urban Studies and Planning (MIT)	General support for the professor, staff, and students
US NSF	United States National Science Foundation	three-year collaborative grant

Support to Chinese Academy of Sciences Team:

UNIDO	United Nations Industrial Development Organization	Support for the 2003 survey
NNSFC	National Natural Science Foundation of China	Support for the CAS faculty and students
Shanxi TVE Bureau	Shanxi Township and Village Enterprise Bureau	Conducted official survey in 2003 of TVE Coke Plants

Other Support:

Swiss Federal Institutes of Technology, Zurich (ETHz)	General support for the professors and staff
The University of Tokyo	General support for the professors and students

Needless to say, this is an extensive list. Multidisciplinary research by a cross-country team of faculty and students is important to conduct in our rapidly evolving world. We learned a tremendous amount not only about the cokemaking sector in Shanxi Province, but also about how people from different cultures react to environmental pollution and how they view and solve the overriding issue of sustainable development. In this book, we obviously can present only a small part of what we have learned, but we hope each reader will enjoy the journey we take you on through the various chapters.

Many thanks to everyone who helped make this book possible, including my many friends and relatives, most of whom still do not understand why I find coke such a fascinating topic on which to spend so many hours.

Karen R. Polenske
May 30, 2005
Cambridge, Massachusetts

FIGURES

TABLES

About The Authors

Karen R. Polenske is professor of regional political economy and planning in the department of urban studies and planning at the Massachusetts Institute of Technology (MIT), where she is head of the International Development and Regional Planning Group and director of the Multi-Regional Planning team. Professor Karen R. Polenske has been at MIT since 1972 and holds a Ph.D. in economics from Harvard University. She is nationally and internationally recognized for her excellence in regional economic research and teaching. She is past President of the International Input-Output Association and won the 1996 North American Regional Science Distinguished Scholar Award and the 1999 Associated Collegiate Schools of Planning Margarita McCoy Award for outstanding service. Her publications include six published books, including *Chinese Economic Planning and Input-Output Analysis* (coedited with Chen Xikang), two forthcoming books—this one on the cokemaking sector in China and the second on the *Economics of the Geography of Innovation*, and numerous articles in key economic and planning journals. Currently, Professor Polenske is analyzing the energy-efficiency of alternative coke and steel industrial technology options in the People's Republic of China (China) and determining socioeconomic effects of land recycling in the North of China. She is team leader for the 15 faculty and students who conducted the Alliance for Global Sustainability (AGS) energy-efficiency research in China discussed in this book.
krp@mit.edu

János M. Beér is professor emeritus of chemical and fuel engineering at MIT. His areas of expertise include air-pollution control, combustion, furnace design, radiative heat transfer, and applied fluid dynamics and chemical kinetics. At the first major US-China joint conference on Clean Coal Utilization, sponsored by the US National Academy of Engineering, he led the US delegation and was co-chairman with Feng Junkai of Tsinghua University. In 1995, he was presented The Axel Axelson Johnson Medal of the Royal Swedish Academy of Engineering Sciences by H.M. Carl XIV Gustaf, King of Sweden, and in January 2004, he was awarded the US DOE's Homer H. Lowry Award for his research on clean utilization of fossil fuels.
jmbeer@mit.edu

Hoi-Yan Erica Chan completed her master's of city planning at MIT in June 2002. She is a former member of the multiregional planning team and focused her attention on the socioeconomic impacts of the evolving cokemaking industry in Shanxi Province. She is currently Consultant, Business Strategy and Policy Group, Steer Davies Gleave, London, England.
echan@alum.mit.edu

Hao Chen received his master's degree from MIT's Technology and Policy Program in June 2000. He is a former member of the multiregional planning team, with research interests in different cokemaking technologies. Currently, he is Developer, IntraLinks, Inc., Boston MA.
chenhao@alum.mit.edu

Xikang Chen is a professor at the Institute of Systems Science, Academy of Mathematics and Systems Science, Chinese Academy of Sciences. He conducts research on input-output analyses, forecasting, and resource economics. He has developed an extended input-output table, which provides considerable detail for the land, labor, and capital inputs, and recently developed a TVE input-output table for China. Additional current research interests include agriculture in the 21st century and the prediction of grain yields in China; working with computable general equilibrium and other econometric models; disaster-assessment techniques related to meteorological factors; and environmental pollution caused by energy consumption in TVEs in China.
xkchen@mail.iss.ac.cn, chenxikang@hotmail.com

Yan Chen received a dual master's degree from the Department of Urban Studies and Planning and the Center for Real Estate at MIT in June 2003. She is a former member of the multiregional planning team, with particular interest in using GIS systems to understand the environmental and cost impacts of coke plant restructuring and relocation in Shanxi Province. Currently, she is Associate, Morgan Stanley.
yanchen@alum.mit.edu

Jinghua Fang was a professor and head of the Department of Thermal Engineering at the Taiyuan University of Technology, Shanxi Province until his retirement in the spring of 2005. He is now consulting in Shanghai. For 15 years, Professor Fang worked at a boiler plant as an engineer and chief engineer, where his two new designs of industrial boilers were awarded second prize by provincial government. After becoming a professor, he spent three years at MIT working with Professor János M. Beér on combustion. In China, he conducts research on coal combustion, energy efficiency, and their related environmental impacts.
fjh@tyut.edu.cn

Wei Guo received her dual master's degree from MIT's Technology and Policy Program and Department of Civil and Environmental Engineering in February 2000. She is a former member of the multiregional planning team with an area of interest in energy-intensity changes in China.
guowei@alum.mit.edu

Holly Krambeck will receive her dual master's degree from the Department of Urban Studies and Planning and Civil and Environmental Engineering (transportation) at MIT in January 2006. She is a member of the multiregional planning team and has been examining the socioeconomic impacts of industrial restructuring in Shanxi Province's cokemaking sector. Other research projects focus on design and implementation of integrated land use and transportation plans in small and mid-size Chinese cities.
krambeck@mit.edu

Jinghua Li is an associate professor, School of Business, China University of Political Science and Law. He received his Ph.D from the Chinese Academy of Sciences in 2004. He worked with Professor Xikang Chen and Dr. Cuihong Yang on the surveys. His current interests involve econometrics, operations research and input-output analysis.
jing-hua_li@163.com, jinghli@tom.com

Uli Matter is Research Scientist at Matter Engineering, Wohlen, which he founded in 2001. Dr. Matter played a significant role with the particulate sensors used by the China Cokemaking Team by overseeing the repair and maintenance of the different sets of sensors.
umatter@matter-engineering.com

Ali Shirvani-Mahdavi received his Ph.D. from MIT's Department of Urban Studies and Planning in January 2005. He is a former member of multiregional planning team and is interested in using input-output techniques to understand improvements in energy efficiency in China.
mahdavi@alum.mit.edu

Masayoshi Sadakata, as of the spring of 2005, is in the Department of Environmental Chemical Engineering, Kogakuin University. He was a professor in the Department of Chemical Systems Engineering at the University of Tokyo during the AGS project. Previous to that, he was a Visiting Researcher in the Department of Chemical Engineering and Fuel Technology at the University of Sheffield, and associate professor in the Department of Chemical Engineering at Gumma University. He is currently conducting research on low-cost approaches to reducing sulfur oxides and on potential improvements to the energy efficiency of coal combustion.
sadakata@chemsys.t.u-tokyo.ac.jp

Steven B. Kraines is an associate professor, Department of Frontier Sciences and Science Integration, the University of Tokyo. His current interests involve applying distributed information technologies to the applied fields of chemical engineering, in particular, developing a modular software platform centered on a dynamic input-output matrix for modeling material and energy cycling in large cities and regions.
steven@prosys.t.u-tokyo.ac.jp

Hans C. Siegmann is with the Center for the Study of Ultrafast Phenomena at Stanford University. Previously, he was professor of physics at the Swiss Federal Institutes of Technology Zurich, where, at his Laboratory for Combustion Aerosols and Suspended Particles, he developed the particle sensors used in the research reported in this book. Those sensors can selectively detect particles based on their surface chemical composition. The sensors, based on his principles, make it possible to monitor respiratory, combustion-generated particles that are relevant to public health issues.
siegmann@slac.stanford.edu

Zhiqiang Qian was a senior research scientist with the Laboratory for Solid State Physics, Swiss Federal Institute of Technology, Zurich, and is currently working in Beijing. He is a physicist specializing in air-pollution monitoring and is helping to design some of the first indoor, outdoor, and personal exposure tests with mobile

sensors detecting physical parameters such as size as well as bulk and surface chemistry. The sensors are specifically designed to cover the respiratory-size ranges of particles relevant for public health.
qianzhiqiang@sohu.com; zhiqiangqian@163.com

Cuihong Yang is an associate professor with the Department of Operations Research and Management at the Institute of Systems Science, Academy of Mathematics and Systems Science, Chinese Academy of Sciences, where she received her Ph.D in 1999. She helped lead the design and implementation of the cokemaking surveys, which laid the foundation for the research presented in this book. She has been conducting numerous economic analyses of the industrial structure in Shanxi Province and China.
chyang@mail.iss.ac.cn

CHAPTER 1

INTRODUCTION

Karen R. POLENSKE[1]

1.0 Coke

Coke is a critical input in the making of iron and steel, thus in the production of major industrial, commercial, and consumer products used in modern society, such as skyscrapers, bridges, industrial equipment, weapons, automobiles, and many consumer appliances. Changes in the coke sector directly or indirectly affect nearly every conceivable measure of the gross domestic product (GDP). Why have coke prices doubled and even more than tripled since 2000? What effect, if any, do coal-supply bottlenecks have on the coke sector in the People's Republic of China (China)? Does the shortage of thermal coal that is used primarily for electricity generation affect the availability of coking (metallurgical) coal that is used for making coke? How do these changes affect people living in China, India, Japan, the United States, and many other developing and developed countries? The answers to these and many other questions may be found only through an understanding of the omnipresent intermediary, the coke sector.

The coke we discuss in this book is a product of the burning of coal or coal blends at temperatures from 900 to 1400 degrees centigrade in a process called carbonization of coal. Coke that is made from metallurgical coal is called metallurgical coke. It is light, yet very strong, and one of its main purposes is to support the heavy iron-ore burden in steel blast furnaces without disintegrating. Although there is both metallurgical coke and foundry coke, we concentrate mainly on metallurgical coke unless otherwise indicated. One reason for this focus is that, of the approximate 130 million metric tonnes of coke produced in China in 2001, more than 80 percent was metallurgical coke (China Coking Technology Congress, accessed June 2004, http://www.mc-ccpit.com/cc/2004/eindex.htm). Most important, many statistical publications do not separate the two types of coke, so that we cannot get sufficient information to discuss foundry coke in detail here.

Karen R. Polenske (ed.), The Technology-Energy-Environmental-Health (TEEH) Chain in China: A Case Study of Cokemaking, 1–8.

A defining characteristic of various coke-oven technologies throughout the world is that the coke ovens are energy intensive and extremely polluting, although some that are beginning to operate seem to be reducing the emissions of most pollutants.

1.1 Background: The Cokemaking Sector

A remarkable phenomenon has been occurring in the People's Republic of China since 1978. From 1978 to 1998, China was able to reduce energy intensity (energy consumption per unit of output) by more than 50%, while other countries at similar stages of development had been unable to accomplish this feat. We began studying the cokemaking sector in 1997 to help determine why this was occurring in China. For example, during this same time period, energy intensities in India increased slightly, while in Iran, intensities increased three-fold. (Polenske 2003). Since national statistics are not always particularly revealing, especially considering China's size and diversity, we examined whether there were particular regions or sectors that were responsible for China's remarkable energy-intensity improvements uncovered by Lin and Polenske (1995) in the early 1990s. We found that for industrial sectors, technological changes were prime contributors to this decrease and that the energy intensity was decreasing in all sectors at the national level. Even so, some regions, such as Shanxi Province, had energy intensities that were twice the level in the country as a whole.

We decided to focus on the cokemaking sector, because it is one of the largest and most energy-intensive industries in the country. From 1980-1997, China became the largest coke producer in the world, producing almost 140 million metric tonnes (hereafter tonnes) of coke (25% of world production) per year by 1997, the year we began our study (Ministry of Coal Industry, 1998), which is probably an underestimate for China, because most so-called "indigenous" coke produced in China may not be included in the production number. Until recently, all coke plants in China have used extremely energy-intensive and highly polluting technologies.

We chose to center our initial analyses on Shanxi Province, because it is the largest coal- and coke-producing region in China (Photo 1). The region produces more than 25% of the total coal and 50% of the coke produced in China. Coal still represents about 75% of the total supply of energy in China, and coke is the largest intermediate coal-consuming sector in Shanxi Province, although in the rest of China it ranks only second or third as a coal consumer. (Polenske and McMichael 2002) Since the 1980s, township and village enterprises (TVEs) have been the main producers of coke in China, producing over 50% of the coke in the country in 1997, up from 20% in 1980, with the remainder produced by state-owned enterprises (SOEs) (UNIDO 2004). Shanxi Province's share of TVE coke production has increased dramatically during the past 20 years, growing from 50% of the total coke produced in the Province in 1980 to over 80% by 1997. (Polenske and McMichael 2002)

Our research led us down a truly fascinating path–not only have we examined how the cokemaking sector has affected national and regional energy intensities, but also how technological changes in this sector and the ongoing shift from production by SOEs to

TVEs are affecting the environment, public health, and regional and national economies. We found that TVEs most recently use technologies that are more profitable, employ more workers, and require lower amounts of investment than SOEs, while SOEs use technologies that are more energy efficient, employ fewer workers, and require larger investments than TVEs. In Chapter 5, Mahdavi discusses in detail this dichotomy in economic profitability and energy efficiency between the SOEs and TVEs.

We started our project with a focus on the coke plants and the coke ovens. During our first field trips I personally used one of the three particulate-pollution sensors — the one that measures polyaromatic hydrocarbons (PAHs). It was on our first field trip to coke plants in 1998 that, although I was riding in an air-conditioned car, I noted extremely high particulate readings from the PAH sensor as we traveled the 150 kilometers of highway along the Fen River from Taiyuan to Hongtong. I said to the deputy director of the Shanxi Environmental Protection Bureau in whose car I was riding, that "The Environmental Protection Bureau could close all the most polluting plants in Shanxi Province, but if they maintain the same coke production by expanding other plants, you will not solve the pollution problem unless you also get improved technology for the diesel trucks." This experience was partially responsible for our incorporation of truck and rail transport into the project.

In Chapter 6, we examine the location of coal mines, coke plants, and the truck and rail transport used to transport the coal and coke and three possibilities for resiting coke plants to reduce the transportation pollution. For our first survey of TVE coke plants in 1998, described in Chapter 2, we did not include questions concerning the location of the coal mines and the method of shipping the coal and coke, but we did include these and related transportation questions in subsequent surveys.

To conduct the extensive analyses of a single sector, coke, I assembled a group of specialists from different disciplines, including chemical engineers, economists, physicists, and planners. Our AGS team members are from the Chinese Academy of Sciences (CAS), Massachusetts Institute of Technology (MIT), Swiss Federal Institutes of Technology, Zürich (ETHz), the University of Tokyo (UT), and the Taiyuan University of Technology (TUT), and we developed close relationships with various officials in the Shanxi Environmental Planning Bureau and the Shanxi Statistical Office, in Shanxi Province, and the National Environment Planning Agency, State Information Office, Ministry of Agriculture, and Agenda 21 offices in Beijing, and with other national and provincial agencies, as well as with several of the major coke plant managers in Shanxi Province.

1.2 The TEEH Chain

As noted earlier, in initial studies to find the causes for the dramatic reduction in energy intensity in China, Lin and I determined that technology played a major role. I therefore developed the TEEH chain concept in 1998 (MRP 1998). (Figure 1.1) One of my major concerns for a number of years has been the health effects of pollution. I am a political economist, not an epidemiologist, but when we started this project, I began to read articles in epidemiology and other literature concerning the health effects of changes in technology. I soon realized that to study the interactions among technology changes, energy, environment, and health, we would need to examine each link in the TEEH chain. Changes in technology (T) are linked to changes in energy use (E), which link to environmental-pollution factors (E), which link to the health (H) of humans. For the cokemaking sector, the technology changes may occur in the SOEs, TVEs, or households (HH). In the latter case, many of the cokemaking households in China still burn coal for cooking and heating, so that new stoves in the households are also needed. Although we did not study this latter technology, we did do some pollution monitoring in the coke-workers' homes.

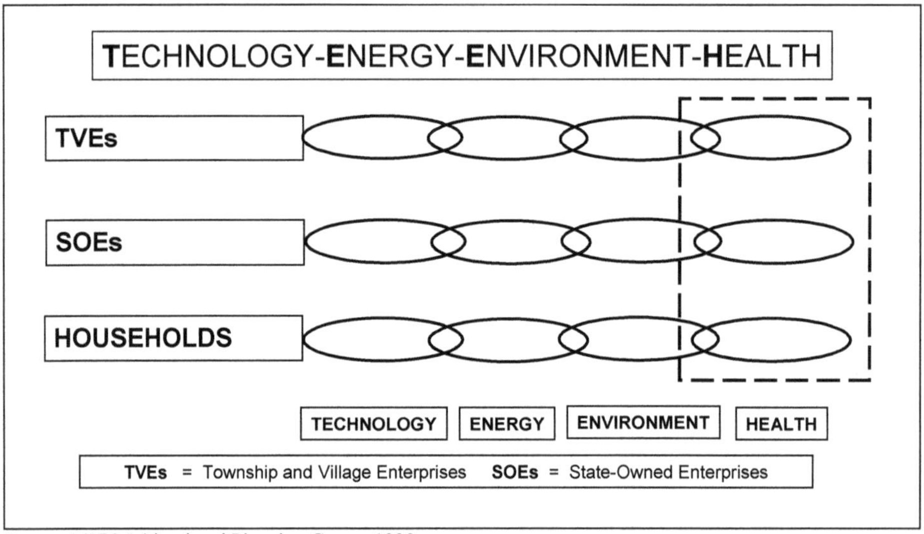

Source: MIT Multiregional Planning Group, 1998.

Figure 1.1: TEEH Chain in China

Also, members of our cokemaking team have observed, how pollution from the production of coke at the plants and from transporting the coal to the plants and the coke to the iron and steel plants is creating serious health problems not only for the coke workers, but also for the people living in the towns and villages near coke plants. In our cokemaking study, we examine each of the first three links (technology, energy, and environment) in depth in relation to the coke sector (Chapters 2-8), and in addition to summarizing our findings, I outline in Chapter 9 how we can study the fourth link, health, when we locate additional funding.

1.3 Research Methods

To conduct such an extensive multidisciplinary study, we used the following methods of economic, environmental, and transportation spatial analysis.

- Plant surveys to obtain energy-intensity and related information for coke plants in Shanxi Province.

- Case studies of and interviews at particular coke plants to collect first-hand information to use as cross-checks on the survey data.
- A geographic information system (GIS) model we developed that incorporates the survey data to help analyze rail and truck shipments of coal and coke along the cokemaking supply chain.
- An extended input-output table (EIOT) for two years at a provincial level to analyze the dramatic changes now occurring in the energy-intensive sectors in China.
- Structural decomposition analysis (SDA) models to analyze the rapid changes in the industrial energy intensities and structures of production, with a spatial extension to include regional energy intensities in a spatial structural decomposition analysis (SSDA).
- Pollution sensors to monitor the ultra-fine particulate pollution from the coke plants, trucks, and households in the cokemaking production areas.
- Interviews during field trips of plant managers, government officials, and others to obtain first-hand knowledge of coal and coke production, transportation, and pollution activities in the region.

For the different surveys, we obtained approval both from the MIT Committee on the Use of Humans as Experimental Subjects (COUHES) and from the relevant provincial agency in China. Where relevant, the authors discuss these methods of analysis in the following chapters.

Throughout the research, we have tried to be consistent in the use of terminology. For example, for weights we use a metric tonne or just tonne, which is 2,200 pounds, versus the short ton, which is 2,000 pounds. Also, we use the Chinese names for the coke ovens, rather than the names more commonly used in the United States and Europe. As an example, we use “modified-indigenous” instead of “beehive,” and “machinery,” instead of “slot” oven. For the oven acronyms, we have a list of all acronyms at the end of the book, so that we do not always identify the oven acronym in the text, partially because some of the explanations are extremely long.

We use Renminbi (RMB) as the unit to express money values. For the entire period of our research project (1997-2004), 8.27 Rmb equals approximately $1, with some slight variations. (U.S. Federal Reserve Bank, accessed June 3, 2005, http://www.federalreserve.gov/releases/H10/hist/dat00_ch.txt; U.S. Federal Reserve Bank, accessed June 3, 2005, http://www.federalreserve.gov/releases/H10/hist/dat96_ch.htm). In China, analysts typically use 10,000 units, rather than 1,000, 100,000, etc. We convert most tables into units of million metric tonnes, or million RMB, for ease in reading by Western audiences.

For the references, we put the Chinese names as in China with the last name first, unless the person lists their name the nonChinese way with first name first.

1.4 Book Overview

Chapter 2 "The Coke Industry in China." Chen Xikang, Yang Cuihong, and Li Jinghua provide a summary of the eight surveys the AGS team has conducted, starting with the 1998 survey of TVE coke plants and local officials and ending with the 2003 TVE survey--all in Shanxi Province.

Chapter 3 "Alternative Cokemaking Technologies in Shanxi Province." Chen Hao and Karen R. Polenske discuss the alternative cokemaking technologies, ranging from the primitive indigenous coke ovens to the modern so-called "clean" ovens. Their discussion is based upon numerous field trips since 1998 to Shanxi Province, each of which involved coke-plant visits—more than 30 plants in all; review of relevant Chinese and English literature about the technologies and changes being made, including technical books, news articles, unpublished papers, and material on relevant websites; participation at five cokemaking conferences in the United States and China; in-depth case studies of three cokemaking plants conducted by Chen Hao (2000); and information provided by government officials, academic colleagues, and plant officials. We recently added to the chapter information on the newest clean-oven technologies that are being introduced in many locations in Shanxi Province.

Chapter 4 "Structural Decomposition Analysis." Ali Shirvani-Mahdavi, Guo Wei, and Karen R. Polenske discuss the coke sector by using extended input-output tables to examine the ways in which the structure of production in Shanxi Province in 1995 varies from that in China in terms of energy use. We also analyze the production changes that have occurred in Shanxi Province from 1992 to 1999, both analyses being based upon available input-output tables. These production changes have helped lead to decreases in energy intensity both in Shanxi Province and in China. The authors show how the main factor affecting these changes in the cokemaking sector is technological change rather than changes in demand by final users or shifts from heavy to light industry.

Chapter 5 "Energy Efficiency and Profitability Differences: State-Owned Versus Township and Village Enterprises." Ali Shirvani-Mahdavi uses input-output data for Shanxi Province and China to provide insights into the differences in the use of energy and in profitability in the SOEs and TVEs, documenting how the SOEs are not profitable, but are environmentally efficient (pollution/unit of output), whereas most TVEs are profitable. The pollution from the TVEs is well known, which other members of the team discuss in detail in Chapter 7.

Chapter 6 "Modeling Cost and Pollution of Coal and Coke Transportation in Shanxi Province." Chen Yan, Steven Kraines, and Karen R. Polenske describe a unique analysis they conducted of the current and potential coke-plant sites and how changes in the size of plants and siting of the plants would affect the amount of coal and coke transportation, hence pollution from the rail and truck transportation. Our discussion of the cokemaking supply chain helps to show that any supply-chain management analyses of automobile, consumer durables, and heavy equipment must include coke plants as a key link in the supply chain.

Chapter 7 "Human Exposure to Ultra-Fine Particulates in the Coke-Making industry in Shanxi Province." In an effort to obtain information to help assess the health effects of cokemaking, Qian Zhiqiang, Hans C. Siegmann, Fang Jinghua, Uli Matter, and Karen R. Polenske provide details on the extensive monitoring for ultra-fine particulates conducted at the coke plants in Shanxi Province. Our focus is on particulates, especially ultra-fine particulates that get embedded into the lungs, because of their importance to the health of the coke workers and the communities in which the workers live. In future work, we anticipate extending the analysis to other forms of air pollution and to water and land pollution—in the latter case to examine the possible remediation of the land where the closed coke plants were.

Chapter 8 "Health of a Region: A Socioeconomic Perspective." Hoi-Yan Erica Chan and Holly Krambeck discuss the socioeconomic effects of cokemaking, based upon information provided in the cokemaking surveys as well as information they collected from their field trips concerning the workers education, training, and safety.

Chapter 9 "Conclusion." I summarize the seven-year study, what the team has accomplished, and the potential for extending the study to examine the health of the workers and people in the communities near to coke plants. From our pollution studies, reported in Chapter 7, we know that severe ultrafine particulate pollution occurs both from the transport of the coal and coke and from the coke ovens and related processes. I believe that pollution from both sources may be causing serious health problems in the towns and villages near to the coke plants. In addition, I discuss the possibility of examining the ways in which coke managers and others in the region may be able to clean up the sites of the numerous closed coke plants for industrial and other uses, turning these sites into useable land for development. In Chapter 9, I review some ways in which the cokemaking team, or others, could conduct studies of these problems.

Thus, in the remaining chapters of this book, the China Cokemaking Team provide the details that we have learned during our seven-year project in Shanxi Province. We are anxious to have others read about the tremendously important changes occurring in its cokemaking sector, the problems being created, and the innovative solutions being proposed by local officials, plant managers, and others to overcome at least some of energy use, pollution-generation, and related problems.

References

Chen, Hao. 2000. Technological Evaluation and Policy Analysis for Cokemaking: A Case Study of Cokemaking Plants in Shanxi Province, China. Cambridge, MA: Massachusetts Institute of Technology, Engineering Systems Division, Master of Science thesis.

Lin, Xiannuan, and Karen R. Polenske. 1995. Input-Output Anatomy of China's Energy-Use Changes in the 1980s. *Economic Systems Research*, 7(1): 67-84.

Ministry of Coal Industry. 1998. *China Coal-Industry Yearbook*. Beijing, China: China Coal Industry Publishing House.

MRP (Multiregional Planning) Group. 1998. Field Trip Notes for Shanxi Province, China, (August). Cambridge, MA: Massachusetts Institute of Technology, Department of Urban Studies and Planning.

Polenske, Karen R. 2003. Environmental Impacts of Energy-Efficient Technologies: Potential for Input-Output and Supply-Chain Analyses in Iran, *Quarterly Iranian Economic Research*, Special Issue on the

Second Input-Output Conference. **14**:1-26.

Polenske, Karen R., and Francis C. McMichael. 2002. A Chinese Cokemaking Process-Flow Model for Energy and Environmental Analyses. *Energy Policy*, **30**(10): 865-883.

UNIDO (United Nations Industrial Development Organization). 2004. Provision of Services for the Execution of a Coking Subsector Survey. UNIDO Contract No. 03/120. Energy Conservation and Greenhouse Gas Emissions Reduction in Chinese Township and Village Enterprises--Phase II. Beijing: UNIDO.

Web Sites:

China Coking Technology Congress, accessed June 2005, http://www.mc-ccpit.com/cc/2004/eindex.htm

U.S. Federal Reserve Bank, accessed June 3, 2005, http://www.federalreserve.gov/releases/H10/hist/dat00_ch.txt

U.S. Federal Reserve Bank, accessed June 3, 2005, http://www.federalreserve.gov/releases/H10/hist/dat96_ch.htm.

---

[1] Professor of Regional Political Economy and Planning. Head, China Cokemaking Team, Department of Urban Studies and Planning, Massachusetts Institute of Technology, USA.

CHAPTER 2

THE COKE INDUSTRY IN CHINA

YANG Cuihong,[1] CHEN Xikang,[2] and LI Jinghua[3]

2.0 Introduction

Cokemaking is the second biggest intermediate consumer of coal in the People's Republic of China (China). Of the total national coal consumption, cokemaking accounted for 13.3% (182.1 million metric tonnes) in 2002. Hereafter, we refer to "metric tonnes" just as "tonnes." The biggest intermediate consumer of coal in China is power generation, which consumed 45.7% (656 million tonnes) of total 2002 coal consumption, but electricity utilies use thermal, not coking, coal. (SSB 2004: 278)

Coking (metallurgical) coal is the most important input into cokemaking. Fifty percent of China's coking-coal reserves are in Shanxi Province. Of the total 2002 coal consumption of 165.9 million tonnes in Shanxi Province, over 51% (85.0 million tonnes) was used for cokemaking and only about 25% (41.7 million tonnes) for electricity generation (*Shanxi Statistical Yearbook* 2003:151). This is the reverse of the two major coal consumers nationally.

In 2003, Shanxi Province produced 72 million tonnes of coke, accounting for about 40% of China's total coke output. Of this total production, Shanxi Province exported 11.6 million tonnes of coke (Shanxi Provincial People's Government accessed 04-24-05, http://www.shanxigov.cn/gb/zgsx/sq/jjgk/dwmy/), accounting for 81% of China's total 13.6 million tonnes of coke exports and 48% of the total global coke supply. Cokemaking has been a major revenue resource for local governments in Shanxi Province, with about 17% of the value added tax and 33% of the revenue tax coming from the cokemaking sector. Because of these significant revenue contributions, energy inefficiencies and pollution stemming from cokemaking and energy policies in Shanxi Province directly influence the province as well as China as a whole. (UNIDO 2004)

Most of our analysis in this chapter relates to township-and-village enterprises (TVEs),

Karen R. Polenske (ed.), The Technology-Energy-Environmental-Health (TEEH) Chain in China: A Case Study of Cokemaking, 9–22.

which since the 1980s, have been the main producers of coke in Shanxi Province. In the early 2000s, they accounted for 85-90% of coke production. In 2003, for example, 63.4 million tonnes of coke were produced by TVE cokemaking plants, comprising about 88% of total coke output in Shanxi Province. However, because of limited funding and obsolete equipment, the energy intensity (in this case, measured as tonnes of coal input per tonne of coke output) of cokemaking TVEs, which are mainly small plants, is high compared to that of larger state-owned enterprises (SOEs). Also, partially because TVEs do not typically recycle the by-products of cokemaking, such as coal-gas and coal-tar, the TVEs usually generate more pollution than their SOE counterparts. (UNIDO 2004)

Since 1995, following national regulations, the Shanxi Provincial Government began to close all indigenous cokemaking TVEs, which use more energy and pollute more than other types of coke plants (TVE and nonTVE). The coal-intensity ratio, for example, for TVEs typically ranges from 1.2 to 1.9 tonnes of coal per tonne of coke, while for SOEs, it is 1.1 to 1.3. The Shanxi government officials had planned to close most of the modified and small-machinery cokemaking TVEs by the end of 1999, so as to establish a "Clean Energy Region of Shanxi," based on large-machinery cokemaking enterprises. Probably because the regulations were not strictly enforced, a few modified and many small-machinery plants remained open. In 2000, China's State Environmental Protection Agency, together with other relevant ministries, took more restrictive measures than earlier for the cokemaking sector, ordering that all indigenous coke ovens and most modified-indigenous coke ovens must be closed by mid-2000. The Shanxi cokemaking industry is in a state of rapid change, trying to restructure while simultaneously trying to respond to the tremendous domestic and international increases in demand for coke and the changes in environmental regulations. (MRP 2001)

Because of the lack of information and inconsistency between different published data sources, our Cokemaking Team placed top priority on obtaining data by surveying cokemaking TVEs to determine the technology, production scale, and regional allocation, and, for comparison, we also conducted surveys of cokemaking SOEs in Shanxi Province. We used these first-hand data on energy use and pollution generated by alternative technologies and the methods we designed to study these data in order to form the basis of our Technology-Energy-Environment-Health (TEEH) chain study, a concept Polenske first presented in a lecture on the 1998 field trip (MRP 1998).

Since 1998, we have conducted eight sample surveys in Shanxi Province: three TVE plant-manager surveys in 1998, 2000, and 2002; two SOE plant-manager surveys in 1999 and 2001; and three local-official surveys for TVEs in 1998, 2000, and 2002. In the fall of 2003, our staff at the Academy of Mathematics and Systems Science, the Chinese Academy of Sciences (CAS), and the Shanxi TVE Bureau conducted a new sample survey of TVEs for a cokemaking study for the United Nations Industrial Development Organization (UNIDO). We first present an overview of those surveys; then, we introduce some of the most important survey results related to TVE cokemaking.

2.1 Survey Preparation

Prior to each survey, we conducted extensive training. The Massachusetts Institute of

Technology (MIT) staff spent several days training those Institute of System Sciences (ISS) staff who would be directly in charge of the surveys. Training included: explaining each question on the questionnaire, considering situations the interviewers might encounter when actually conducting the survey, and explaining what to do if the interviewees give ambiguous or wrong information. ISS staff then organized a two-day training session with Shanxi local officials from prefectures, cities, towns, and villages. Depending on the local situation, we usually trained more than 25 local officials from coke-producing regions.

Cokemaking technology in Shanxi Province is extremely varied, with more than fifteen kinds of coke ovens currently (2005) in use, details of which are presented in Chapter 3. Before 2000, there were even more types. The large number of oven types and the fact that some plants have more than one type of oven create difficulties in selecting plants for each survey. To guarantee the integrity of the surveys, we therefore use scientifically designed sampling methods. We adopt a layered, stochastic sampling method, in which oven type, its regional allocation, and coke output form the main layers. For each coke-oven type, we distribute a calculated number of surveys according to the proportion of the total coke plants in Shanxi Province using that type of oven. Thus, before 2000, because modified-indigenous coke ovens were very popular, we include a relatively large number of these plants in our sample for 1998 and 2000, but not for later years.

We have conducted each survey in two phases. First, members of the ISS survey team personally interviewed plant managers and local officials to get the baselines for the survey. Second, because of travel and other complications in remote mountain regions, the ISS team trained local officials in Taiyuan so that they could then conduct the survey and mail survey forms to cokemaking TVEs and local governments.

2.2 General Survey Data

To determine some of the major factors contributing to the key changes in Shanxi's current cokemaking industry, the Cokemaking Team, led by Polenske, designed a detailed survey questionnaire with seven main sections (Table 2.1).

Table 2.1: Coke-Plant-Survey Outline

Survey Section		Description
1	General situation	Oven type and its properties, output, reasons for oven selection, and ownership
2	Byproducts	Total output, recycling of, and related market conditions
3	Coal consumption and coke production	Coal type and volumes, coke type and markets
4	Facility and equipment	Methods and frequency of upgrading equipment, frequency of repairs, and coal-input coefficient of each oven type
5	Financing	Methods plants use to finance their fixed and flow assets, taxes, and fees
6	Employment	Number of employees, structure of employees, training and education level of employees, and safety measures
7	Pollution	Types of pollutants measured

Source: AGS (1998). TVE Cokemaking Survey, 1998.

Starting with the 2001 survey, we added water-use because of water shortages in Shanxi Province. Also, to obtain changes in technology and energy efficiency, we asked plant managers to describe any significant changes since the 2000 survey.

In 1998, we distributed 259 surveys to the TVE plants and 50 to local officials, of which we conducted 29 plant surveys and 10 local official surveys in person. We conducted the others by mailing questionnaires to the plants. Of the 245 total cokemaking TVE plants and 45 local officials who returned the forms, we certified that 158 cokemaking TVEs and 27 local officials (64.5% and 60.0%, respectively) were sufficiently complete to tabulate. These certified plants produced 5.2 million tonnes of coke in 1997, about 11% of the total 1997 TVE coke output in Shanxi Province. Of the TVE plants, some were large-scale (100,000 tonnes or more), but most were medium and small-scale (less than 100,000 tonnes), and some had high and some had low energy-efficiency, with coke-to-coal ratios of 1:6, and/or pollution (high emissions of SO_2 and other gases). (Table 2.2; UNIDO 2004)

In 2000, we surveyed 258 TVE cokemaking plants and 40 local officials, of which the survey team personally interviewed 8 plant managers. Of the 208 cokemaking TVEs and 31 local officials who returned the forms, we certified that 164 cokemaking TVEs and 22 local officials (78.8% and 71.0%, respectively) were sufficiently complete to tabulate. These certified plants produced 11.2 million tonnes of coke in 2000, about 30% of the total 2000 coke produced by Shanxi cokemaking TVEs. (Table 2.2; UNIDO 2004)

In 2002, we surveyed 250 TVE cokemaking plants and 40 local officials. Among the 175 cokemaking TVEs and 37 local officials who returned the questionnaires, we were able to use forms for 127 plants and 35 local officials (72.6% and 94.6%, respectively). These certified plants produced 12.8 million tonnes of coke in 2002, about 24.5% of the total 2002 TVE coke output in Shanxi Province. (Table 2.2; UNIDO 2004)

Table 2.2: Number of Shanxi Province Coke Plants and Surveys, SOE and TVE, 1998-2003

Survey Description	Year	Number of Plants at Time of Survey	Number of Surveys Distributed	Number of Surveys Returned	Number of Useable Surveys	Number of Interviews Made by ISS Staff
TVE plant manager surveys	1998	1,872	259	245	158	29
	2000	1,076	258	208	164	8
	2002	683	250	175	127	4
	2003	589	150	131	110	5
Local official surveys for TVEs	1998	n.a.	50	45	27	10
	2000	n.a.	40	31	22	0
	2002	n.a.	40	37	35	0
	2003	n.a.	30	27	23	0
SOE plant manager surveys	1999	n.a.	8	8	7	8
	2001	n.a.	55	49	49	0

Source: AGS (1998-2002). TVE Cokemaking Survey, 1998-2002.
UNIDO (2004).
n.a. = not applicable or not available.

In order to trace the many changes occurring in the cokemaking industry in Shanxi Province since 1995, we conducted a new TVE survey from October to the end of December 2003. We distributed 150 questionnaires to the plants, and 131 forms were returned. After certification, we determined that 110 questionnaires were sufficiently complete to tabulate.

From the survey and other information, we determined that TVE coke plants produced 33.4 million tonnes in 2003. During the period of the surveys (1998-2003), while the number of coke plants declined from 1872 plants in 1998 to 589 plants in 2003 (Table 2.2), the amount of coke output at TVE plants has remained about the same (decreasing only slightly from 33.8 million tonnes in 1998 to 33.4 million in 2003 (UNIDO Table 4.3). The average increase in output per plant may have been caused by many factors, although one of the most significant factors is the increasing number of large-scale plants.

TVE coke production increased in importance compared with SOE coke production during the time of our surveys. We surveyed the SOEs in 1999 and 2001. Because the 1999 SOE survey was a pilot study, we conducted a limited survey by visiting 8 SOE cokemaking plants and not distributing any survey forms by mail. We did this pilot study in order to gain an overview of Shanxi SOE coke plants, including their technology, management, and quality of employees. In 2001, we surveyed 55 SOE cokemaking plants and received 49 qualified copies after certification. These certified plants produced 8.0 million tonnes of coke in 2001 (Table 2.2; UNIDO 2004).

2.3 Evolution of Shanxi Province Cokemaking

From the surveys, we find that some major improvements have occurred between 1998 and 2003 in TVE cokemaking in Shanxi Province, especially in terms of the type of ovens being used, their improved energy efficiency, and the expansion of the TVE coke plant output, but the level of education and training of the personnel has not changed much, and the number of TVE coke workers decreased, partially as a result of an increase in labor productivity in the TVEs.

2.3.1 Shanxi Cokemaking TVEs

In this section, we describe the evolution of coke-oven technologies used by Shanxi Province TVEs. If the reader is unfamiliar with Chinese coke-oven-naming conventions, we refer the reader to the list of acronyms at the end of this book, as well as Chapter 3, in which Chen and Polenske describe the features of the ovens in detail.

Before 2000, although many cokemaking TVEs constructed small-machinery coke ovens, most used modified-indigenous coke ovens, of which TJ-75 type (Taiyuan Jixiehua 1975), 91 type (1991-type), and SJ-96 (Sanjia-1996) were the most common (Figure 2.1 and Chapter 3). Most cokemaking TVEs are medium- and small-scale (i.e., with an annual coke output of less than 100 thousand tonnes), while large-scale plants with annual coke output over 100 thousand tonnes account for less than 20% of the output. Only a few TVE plants recycle by-products, such as coal-tar and coal-gas. The energy intensity of cokemaking TVEs is relatively high, although recently local officials and plant managers have been trying to improve coke-oven technology. Thus, we find a great potential for Shanxi cokemaking TVEs to save energy and resources and, in the process, to reduce pollution and improve the health of the workers and people in the local community.

On December 31, 1999, a new industrial policy, entitled *The Catalogue for Elimination of Less-Advanced Production Capacity, Technology, and Products (the Second Series)*, was enacted by the State Economic and Trade Committee of China. The Catalogue covered eight industries, of which the first one is iron and steel-related, including indigenous and modified-indigenous, cokemaking technology and coke-oven projects with a battery height less than four meters. (UNIDO 2004) Because Shanxi Province officials began to implement this policy, especially since 2000, Shanxi cokemaking TVEs experienced many technology changes, the most important of which we explain in the following paragraphs.

First, currently (2005), the cokemaking TVE technologies in Shanxi Province have improved significantly since 1998 in terms of their energy efficiency. In 1998, local officials deemed the SJ-96 (Sanjia 1996), introduced by the Shanxi Sanjia Coal Chemistry Co., Ltd., in Jiexiu (Photo 15), the JKH-97 (which were designed by Sansheng Coke Company in Jiexiu City in 1997 and named JKH-97—Photo 9), and small-machinery coke ovens to be the most efficient ovens; whereas, today (2005), the local authorities are promoting the use of clean coke ovens, such as QRD-2000 (Qingjie Rehuishou Daogu Lu--clean oven with tamping and heat recovery) designed in 2000, and the newly designed machinery coke ovens, such as JNK43-98D, designed in 1998, which are 4.3 meters high, with tamping (Daogu), designed by the Anshan Cokemaking & Refractory Engineering Consulting Corporation, and the TJL4350D Taiyuan Jixiehua Lu (mechanized oven that is 4.3 meters high, 500 mm wide, and features coal tamping). Plant managers were forced by local officials to close most indigenous coke ovens. Accordingly, they now widely adopt machinery coke ovens and some advanced modified-indigenous coke ovens, such as SJ-96 with heat recovery (Photo 16). Of the surveyed plants, 24.4% employed machinery coke ovens in 2000, many more than in 1998 (Figure 2.1). Up to 2000, the output of machinery coke ovens surveyed accounted for 41.2% of the total output surveyed. In addition, coke-plant managers began to operate at least three large machinery coke plants, with coke output of more than 600,000 tonnes and with 4.3 meter-high ovens. (UNIDO 2004)

From the 2002 TVE survey, we find that the number of plants that adopted small-machinery coke ovens increased rapidly since 1998, accounting for 47% of the total plants surveyed. Although small-machinery coke ovens do not have advantages over modified-indigenous ovens in terms of environmental pollution, the 1999 industrial policy did not list these ovens as one of the less-advanced technologies that should be eliminated. Another major change in 2002 was the establishment in TVEs of some large-machinery coke ovens, for example, the JNK43-98 ovens. At the same time, new types of ovens, that are locally called “clean” coke ovens, such as DQJ-50, the Daogu Qingjie Lu (tamping clean oven, for 50 tonnes capacity per battery), and the YX-21QJL-1 (Yingxian 21 Shiji Qingjie Lu-1, in Chinese; that is, Yingxian 21 Century Clean Oven-type 1, in English), were built and put into operation in some plants. (UNIDO 2004)

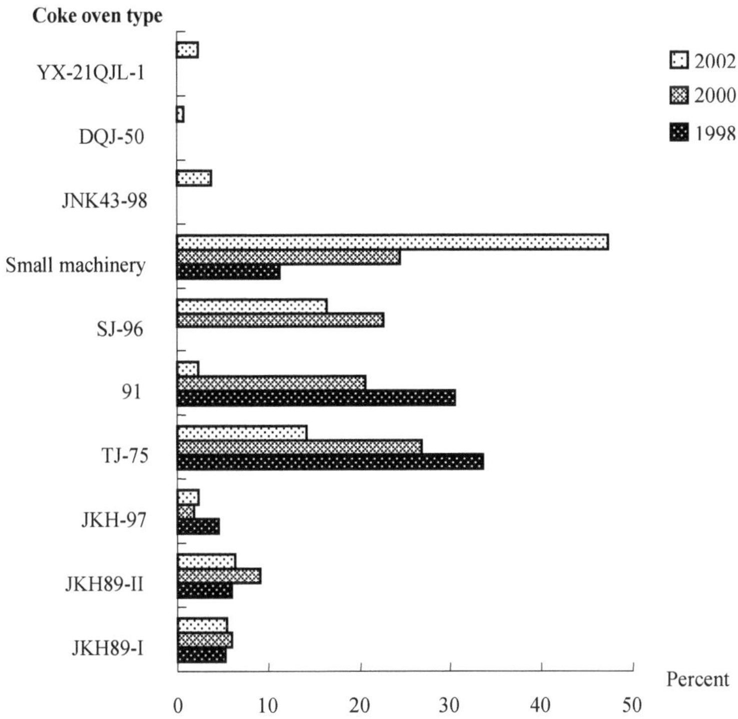

Source: AGS (1998, 2000, and 2002). TVE Cokemaking Survey, 1998, 2000, and 2002.
Note: Oven names are provided in the list of acronyms. The initials usually indicate the city/town/village in which the machinery was designed and the last number is often the year in which the oven was designed. The number on the X axis is the percentage of the total number of each type of coke oven surveyed; while on the Y axis is the name of the coke-oven type. Here, small machinery refers to small-machinery coke ovens, including Red Flag type, small 58 type, and WJ-663 type. We did not subdivide it into specific types, because, in the 1998 survey, we only have total number of plants that adopt small-machinery coke ovens, not the specific type of oven.

Figure 2.1: Shanxi Province Coke-Oven Structure, 1998, 2000, and 2002

In 2003, more Shanxi Province TVE coke-plant managers installed large-machinery coke ovens. Statistics show that 54 newly built coke plants were put into production in China in 2003, with 23.2 million tonnes of new coke capacity, of which Shanxi Province TVEs installed ovens with a coke capacity of 8.6 million tonnes at 24 coke batteries. (China Coking Industry Association—CCIA, accessed 01-18-05, http://www.cnljxh.com/zj-xgc.shtml)

Second, energy efficiency has increased considerably since 1998 (Figure 2.3). TVE coke plants with high coal-input coefficients (tonnes of coal input per tonne of coke output) between 1.4 and 1.6 accounted for 31.1% and 20.5% of the total surveyed in 2000 and 2002, respectively, which is much lower than the 43.0% in the 1998 survey. Of the total

TVE plants surveyed in 2002, 60.6% had a coal-input coefficient of coke less than 1.4, a much higher percentage than the 26.6% in the 1998 TVE survey.

From the 2000 and 2002 TVE surveys, we find that as many as 12.8% and 11.8%, respectively, of the surveyed TVEs consume less than 1.2 tonnes of coal to produce one tonne of coke, which is at the same level as that of large SOEs (Figure 2.3). This decrease in coal intensity, which is the reciprocal of the increase in coal efficiency, is caused by many factors, the most important being improvements in oven technology. In Shanxi Province, the average technology improved in terms of the coal-input coefficient not only by the introduction of improved coke ovens, but also by the closure of ovens with low energy efficiency. Although some indigenous coke ovens can produce coke of fairly good quality, they use considerable coal per unit of coke output; therefore, since 1995 (and especially since 2000), the Shanxi government has closed more and more indigenous coke ovens, thus increasing the energy efficiency of the whole coke sector.

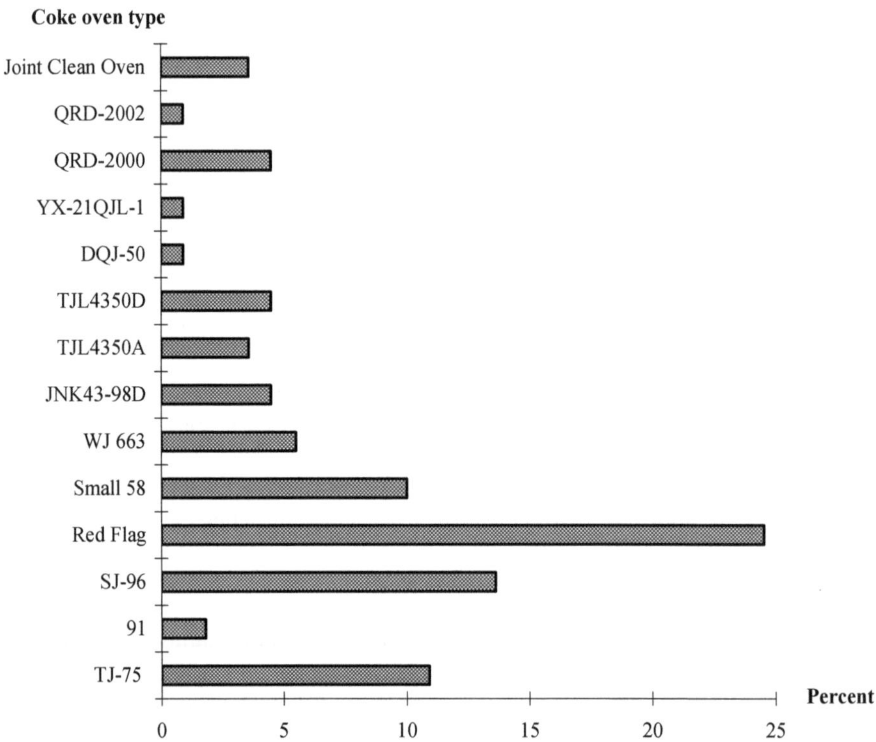

Source: UNIDO (2004).
Note: In this figure, Red Flag, Small 58, and WJ 663 are the small machinery coke ovens mentioned in Figure 2.1. The number on the X axis is the number of each type of coke oven as a percentage of the total number surveyed, that on the Y axis is the name of the coke-oven type.

Figure 2.2: Shanxi Province Coke-Oven Structure, 2003

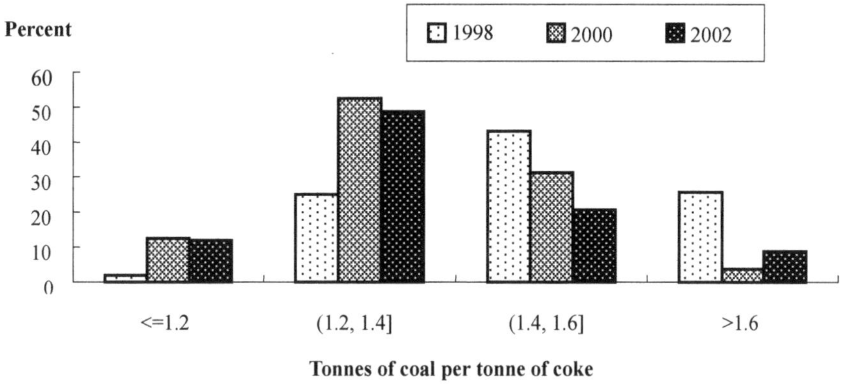

Source: AGS (1998, 2000, and 2002). TVE Cokemaking Survey, 1998, 2000, and 2002.
Note: The number on the Y axis is the percentage of the total number of plants surveyed, while the numbers on the X axis are the coal-per-tonne-of-coke data.

Figure 2.3: Changes in Energy Efficiency in Shanxi Province Cokemaking TVEs, 1998, 2000, and 2002

Third, the production scale of the average plant has increased since 1995, especially after 1999. The total number of TVE cokemaking plants has been decreasing rapidly, but the coke output has remained relatively constant. After the implementation of the 1999 national industrial policy, officials closed many small-scale coke plants. From the 2003 TVE survey, we determined that the number of TVE cokemaking plants changed very rapidly, declining from 1,872 in 1998 to 589 in 2003. According to the survey results, however, total coke TVE output decreased only slightly—from 33.8 million tonnes in 1998 to 33.4 million tonnes in 2003. (Table 2.2; UNIDO 2004) The reported coke output of TVEs in Shanxi Province is 63.4 million tonnes (UNIDO 2004). We note that although the Shanxi TVE Bureau staff intended in their survey to determine the output of all TVE plants in the Province, some plants did not report their information, so that the staff could account for only about half of the total output.

From our 2000 and 2002 TVE surveys, we recorded more large plants with annual coke output above 100 thousand tonnes than in previous years. The 2003 TVE survey shows the same trend of increase in production scale, large plants with annual coke output above 100 thousand tonnes account for 49.1% of the output. (Table 2.3) Up to 2002, however, medium-size and small cokemaking TVEs with annual coke output between 20 to 100 thousand tonnes still account for the greatest proportion (42.5% in 2002) of the number of TVE plants.

Fourth, we find that cokemaking TVEs mainly employ personnel with little formal education. In the surveyed cokemaking TVEs, for example, most employees (more than 50%) have only 6-12 years of school, which is the middle-school level (Table 2.4). Although some of the technicians and administrators are highly educated with over 12 years of schooling, still, many workers, and a few technicians and administrators, have had little education, with fewer than six years of school, which is the primary-school level. Compared with the results in the 1998 and 2000 TVE surveys, this situation changed a lot since 2000 after the coke-oven restructuring policies were implemented by the Shanxi Government. Even so, those employees with fewer than 12 years of education still form a large proportion of the coke employees, and such a coke-employee education level is, and will continue to be, one of the constraints on improving the level of the TVE management and technology, unless some measures are taken to overcome the problem. Plant managers who want to employ advanced technology need highly educated workers in order to make full use of the plant capacity. This is a topic Chan and Krambeck discuss again in Chapter 8.

Table 2.3: Coke Output of Surveyed Shanxi Province Cokemaking TVEs, 1995-2002

Unit: percent

	tonnes			
Year	<20,000	20,000-100,000	100,000-200,000	>=200,000
1995	6.3	26.0	5.5	2.4
1996	10.2	29.1	4.7	3.1
1997	14.2	31.5	6.3	3.9
1998	15.0	31.5	7.9	5.5
1999	14.2	40.9	7.1	7.1
2000	9.4	38.6	11.8	7.1
2001	10.2	50.4	14.2	8.7
2002	7.1	42.5	21.3	12.6

Source: AGS 1998-2002. TVE Cokemaking Survey, 1998-2002.
Note: Some of the plants surveyed did not provide information, so that the percentages do not sum to 100%.

Table 2.4: Shanxi Province TVE Employee Education 1998, 2000, and 2002

Unit: percent

	Years of Education		
Year	Less than 6 years	6-11 years	12-18 years
Production worker			
1998	34.2	46.2	0
2000	32.9	51.8	0.6
2002	28.3	52.8	5.5
Technical staff			
1998	11.4	46.8	28.5
2000	14.0	43.3	26.8
2002	10.2	36.2	37.0
Administrative staff			
1998	12.0	57.0	17.1
2000	12.2	51.2	20.1
2002	7.9	36.2	36.2

Source: AGS (1998-2002). TVE Cokemaking Survey, 1998-2002.
Note: Some of the plants surveyed did not provide information, so that the percentages do not sum to 100%.

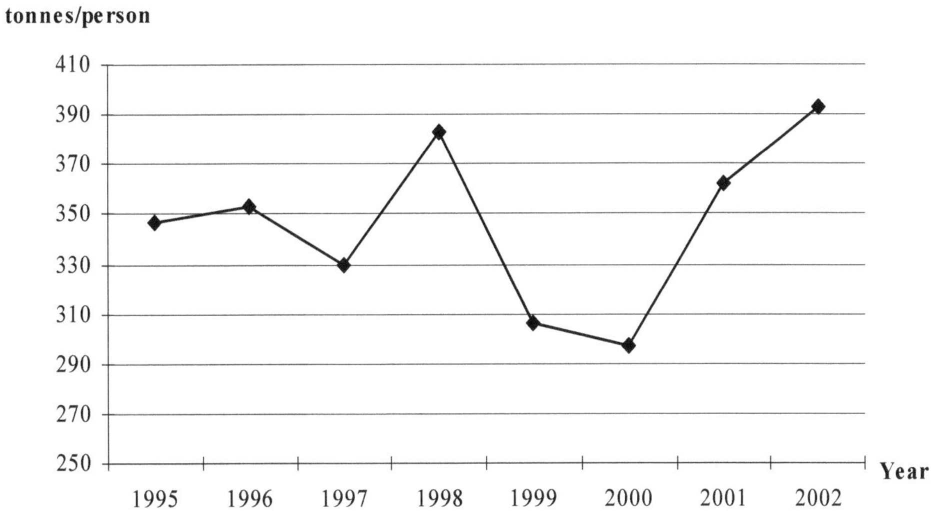

Source: AGS (1998-2002). TVE Cokemaking Survey, 1998-2002.

Figure 2.4: Labor Productivity Trend, Shanxi Province Cokemaking TVEs, 1995-2002

Fifth, the 2002 local official TVE survey shows that the number of TVE employees has decreased by about 1.5% from 85,790 in 1995 to 84,464 persons in 2002, while the coke output increased by 11.5% from 29.8 million tonnes in 1995 to 33.2 million tonnes in 2002. Thus, as the number of TVE employees decreased, the productivity of the remaining employees increased, and the productivity was higher in 2002 than in 1995 (Figure 2.4). In 1995, productivity was 347 tonnes/person (i.e., one TVE employee produced 347 tonnes of coke, on average), while in 2002, productivity rose sharply to 393 tonnes of coke per employee, though with fluctuations during 1998 and 2000.

Of the surveyed cokemaking TVE plants, 43% exported coke in 2000, a 16 percentage point increase from the 27% in 1998. Shanxi TVE cokemaking plants exported 9.5 million tonnes of coke in 2002, the latest year for which data are currently (2005) available, accounting for 70% of China's total coke exports (Stainless Steel Information, Accessed 01-18-05, http://xhhg.51bxg.com/news/lme/200412/161823.html).

TVE managers have been aware of environmental problems, and many coke plant managers are starting to measure pollutants. Currently, the newer TVEs are more energy efficient and less polluting than the older ones, as we discuss below, but they are less energy efficient than the SOEs, as Mahdavi discusses in Chapter 5.

2.3.2 Comparison Between Shanxi Cokemaking SOEs and TVEs

Based upon a careful review of the SOE and TVE survey data, we have the following four major findings.

1. SOEs are far more environmentally/energy efficient (amount of pollution produced/energy used per unit of output) than TVEs, but the TVEs are more profitable (revenue per unit of output). In 2001, 42.9% of surveyed cokemaking SOEs had coal-input coefficients of coke less than 1.3, and nearly 33% of the plants had coefficients between 1.3 and 1.4. Only about 25% of the plants had coefficients over 1.4. Most of the surveyed TVEs had coal-input coefficients of coke between 1.4 and 1.6, and about 25% of surveyed cokemaking TVEs had coal-input coefficients of coke over 1.6. Compared with TVEs, SOEs had relatively low coal-input coefficients of coke and are therefore more energy efficient, thus less energy intensive than TVEs.

2. TVEs use technologies that are more profitable, employ more workers, and require lower capital investment than SOEs. SOEs use technologies that are more energy efficient, thus less energy intensive, employ fewer workers, and require larger investments than TVEs. Machinery coke ovens are the main technology of the surveyed Shanxi cokemaking SOEs. About 66% of the SOEs employ machinery coke ovens, which is more than the TVEs. In 2002, about 51% of the TVEs employed machinery coke ovens, in 2000 only 25%, and even fewer did so in 1998 (Table 2.5).

3. SOE plants are larger in scale than TVE plants, producing more output on average and have more advanced coke ovens and other machinery (Table 2.6). Most of the SOEs are large or medium scale, and over a quarter of them produce more than 200,000 tonnes of coke annually. Large SOE plants with coke output of more than 100,000 tonnes account for 55.1% of total output in Shanxi Province in 2001, and 26.5% of the SOEs have annual coke output of more than 200,000 tonnes. In contrast, most TVE coke plants are small and medium in size, using modified indigenous ovens. Even so, the number of large TVE plants with coke output of more than 100,000 tonnes has been increasing over the past several years, accounting for 33.9% of total surveyed plants in 2002 (15.9% in 2000), compared with only 6.5% of the total in 1995. (AGS 2000). For most of these plant managers, quality and cost are higher priorities for coke-oven selection than the full recovery of environmentally hazardous by-products. (AGS 2000). We can see these differences clearly in Figure 2.5.

   For employees, the same pattern exists as for output, with SOEs having more employees per plant than TVEs. Whereas 51.0% of the surveyed TVE plants had fewer than 200 employees in 2001, most SOEs have more than 200 employees. In total, SOEs accounted for 65.3% of all coke employees in 2001; in fact, the large SOEs with over 1,000 employees accounted for 16.3% of the surveyed SOEs. Thus, SOEs are still relatively labor intensive, partially because of the historical contributions to employee welfare, discussed in more detail in Chapter 8.

4. The quality of employees, measured in terms of schooling, in SOEs is better than that of TVEs. For example, nearly 70% of the SOE technical staff had at least 12 years of schooling in 2001, while only 37% of TVE technical staff had achieved similar

schooling levels in 2002. Also, almost 86% of SOE production workers had 6-12 years of school in 2001, while in TVEs only 52.8% had 6-12 years in 2002. These results are not too surprising, because TVEs are often run by peasant farmers, and most workers come from nearby towns or villages. The low education level of the coke workers may become a great obstacle to increasing competitiveness in the coke market. (AGS 2000)

Table 2.5: Shanxi Province Oven Structure, SOEs (2001) and TVEs (2002)

Unit: percent

Plant Ownership	Oven Structure						
	TJ-75	91	Machinery	SJ-96	JKH89-I	JKH89-II	JKH-97
SOEs	2.9	5.9	66.1	13.2	7.4	4.4	1.5
TVEs	14.2	2.4	51.2	16.5	5.5	6.3	2.4

Source: AGS (2000). SOE Cokemaking Survey, 2001, TVE Cokemaking Survey, 2002.
Note: Some plants have more than one type of oven.

Table 2.6: Shanxi Province Coke Output, SOEs (2001) and TVEs (2002)

Unit: percent

Plant Ownership	tonnes			
	<20,000	20,000-100,000	100,000-200,000	>=200,000
SOEs	4.1	40.8	28.6	26.5
TVEs	7.1	42.5	21.3	12.6

Source: AGS (2000). SOE Cokemaking Survey, 2001, TVE Cokemaking Survey, 2002.
Note: Some of the plants surveyed did not provide information.

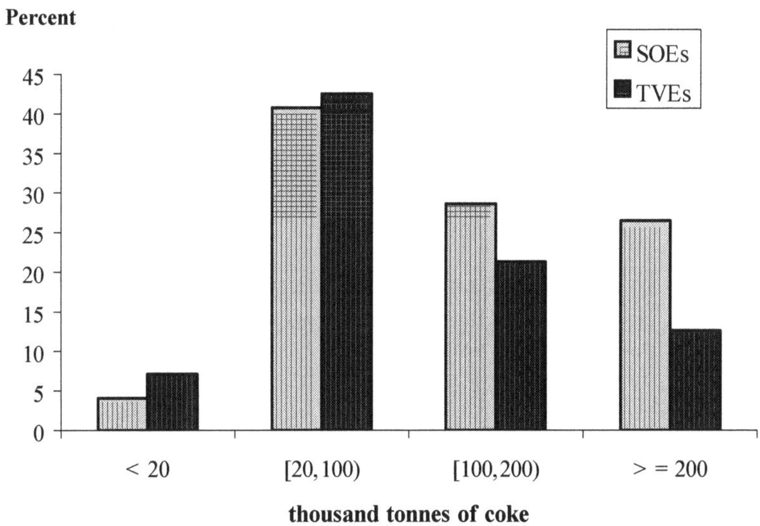

Source: AGS (2000). SOE Cokemaking Survey, 2001, TVE Cokemaking Survey, 2002.

Figure 2.5: Shanxi Province Coke Output, SOE (2001) and TVE (2002)

In conclusion, our analyses of the SOE and TVE survey data indicate that Shanxi Province coke managers face issues concerning energy efficiency, employment, and environmental regulations. The newer TVEs are more energy efficient and less environmentally polluting than the older ones, but still not as energy efficient as the SOEs. Also, the newer TVEs employ far fewer manual workers than the older ones. On the whole, cokemaking SOEs and TVEs are critical both for the local economy in Shanxi Province and the world market. These cokemaking plants are experiencing quick technological transitions, as managers adopt more advanced coke ovens.

References

AGS (Alliance for Global Sustainability). 1998. AGS TVE Cokemaking Survey, 1998. Cambridge, MA: AGS Cokemaking Team, Massachusetts Institute of Technology.

AGS (Alliance for Global Sustainability). 1999. AGS SOE Cokemaking Survey, 1999. Cambridge, MA: AGS Cokemaking Team, Massachusetts Institute of Technology.

AGS (Alliance for Global Sustainability). 2000. AGS TVE Cokemaking Survey, 2000. Cambridge, MA: AGS Cokemaking Team, Massachusetts Institute of Technology.

AGS (Alliance for Global Sustainability). 2001. AGS SOE Cokemaking Survey, 2001. Cambridge, MA: AGS Cokemaking Team, Massachusetts Institute of Technology.

AGS (Alliance for Global Sustainability). 2002. AGS TVE Cokemaking Survey, 2002. Cambridge, MA: AGS Cokemaking Team, Massachusetts Institute of Technology.

MRP (Multiregional Planning) Group. 1998. Field trip Notes for Shanxi Province, China, (August). Cambridge, MA: Massachusetts Institute of Technology, Department of Urban Studies and Planning.

MRP (Multiregional Planning) Group. 2001. Field trip Notes for Shanxi Province, China, (July). Cambridge, MA: Massachusetts Institute of Technology, Department of Urban Studies and Planning.

Shanxi Province Statistical Bureau. 2003. Shanxi Province Statistical Yearbook, 2003. Beijing: China Statistics Press.

SSB (State Statistical Bureau of China). 1998. China Statistical Yearbook, 1998. Beijing: China Statistics Press.

SSB (State Statistical Bureau of China). 2003. China Statistical Yearbook, 2003. Beijing: China Statistics Press.

UNIDO (United Nations Industrial Development Organization). 2004. Provision of Services for the Execution of a Coking Subsector Survey [for 2003]. UNIDO Contract No. 03/120. Energy Conservation and Greenhouse Gas Emissions Reduction in Chinese Township and Village Enterprises--Phase II. Beijing: UNIDO.

Internet Sites:

China Coking Industry Association–CCIA, accessed 01-18-05, http://www.cnljxh.com/zj-xgc.shtml.

Stainless Steel Information, accessed 01-18-05, http://xhhg.51bxg.com/news/lme/200412/161823.html.

---

[1] YANG Cuihong, Associate Professor, Academy of Mathematics and Systems Science, the Chinese Academy of Sciences, Beijing, People's Republic of China.

[2] CHEN Xikang, Professor, Academy of Mathematics and Systems Science, the Chinese Academy of Sciences, Beijing, People's Republic of China.

[3] LI Jinghua, Ph.D. School of Business, China University of Political Science and Law, Beijing, People's Republic of China.

CHAPTER 3

ALTERNATIVE COKEMAKING TECHNOLOGIES IN SHANXI PROVINCE

Hao CHEN[1] and Karen R. POLENSKE[2]

3.0 Introduction

China is using both some of the oldest still-existing cokemaking ovens and also some of the latest, most energy-efficient and least-polluting technologies to make coke from coal. These technologies differ from each other during the several cokemaking stages; consequently, they have diverse environmental and health impacts. In order to differentiate the technologies, we summarize the basic cokemaking concepts and characteristics of the major technologies used in China, especially those employed in Shanxi Province.

Coke is produced in three stages (Table 3.1). During Stage 1, coal is dried at a temperature of 100 to 250°C; in Stage 2, coal is melted until it becomes pliable and sticky (~300°C). At this stage, coal is swelled by internal gases and it becomes porous and highly reactive in the later stages. After the sticky material begins to swell (~350°C), it is heated even further in Stage 3 until it solidifies and shrinks, eventually hardening to semi-coke (400-500°C) and to coke (500-900°C). In this three-stage process, the most important reaction is the thermal decomposition that occurs in Stage 2, which converts the complex molecular structure of coal (basically hydrocarbon) into a simple solid (mainly carbon) and creates various by-products, such as tar, benzene, and coke-oven gas (COG) (Li 1995: 4).

Although cokemaking generally follows the same three-stage process, the detailed production process and related technologies are, to some extent, related to types of coke being produced. Coke is widely used in iron and steel making, non-ferrous metallurgy, metal casting, sintering, gasification, and calcium-carbide making. Slightly different

Karen R. Polenske (ed.), The Technology-Energy-Environmental-Health (TEEH) Chain in China: A Case Study of Cokemaking, 23–40.

Table 3.1: Stages in the Coking Process

Stage	Temperature	Phenomenon
1	<=300°C	Coal dries in the coke oven at 100-250°C; At 300°C, gases trapped in the coal become volatile as the bitumen is decomposed.
2	300–600°C	Coal softens, melts, and becomes plastic and sticky; Volatiles trapped in the coal blow the plastic material into hollow structures; At 400-500°C, the matter starts to solidify, until semi-coke forms.
3	600°C–1000°C	Semi-coke hardens and shrinks to form coke. Methane, hydrogen, and carbon monoxide are also released.

Source: Li, Shaojing and Shen Weiqing (1995: 5), supplemented by comments from Professor Fran McMichael, Spring 2000.

varieties of coke are used for each of these processes. In this chapter, we focus on two particular kinds of coke: metallurgical and foundry coke. About 90 percent of the coke produced in Shanxi Province is metallurgical coke, which is mostly used for blast-furnace iron-making and steel-making (Chen 2000). In blast furnaces, coke is both a fuel and a reduction agent, and it also acts as a skeleton to support the iron-ore bed. To achieve these purposes, the metallurgical coke pieces usually need to be relatively large (from 20 millimeters (mm) to 80 mm per piece), strong, and non-porous, and have low reactivity. Also, good metallurgical coke should have few impurities; particularly, it should be low in ash and sulfur content. (Li 1996; MRP 2000) Foundry coke is used to generate heat for melting metal and reheating fluidized metal. It also acts as a supporting skeleton to ensure good ventilation when making metal castings. Foundry coke is usually stronger than metallurgical coke and possesses fewer impurities.

The quality of coke largely depends on the quality of the coal used to make it. Good-quality-coking coal is usually low in ash and sulfur content and meets certain standards for water content, volatility, and the caking propensity. In China, the best caking coal is called coking coal (Table 3.2). Coal that has poor caking propensities needs to be blended with some coking coal when making coke. Blending coal with different caking propensities and volatilities can reduce dependence on the use of high-quality coal and make the coke products more competitive in terms of price.

3.1 Characteristics of Major Cokemaking Technologies

A cokemaking technology is defined by its coke ovens, auxiliary equipment, and by-product recovery system. We classify the major cokemaking technologies that have recently been available in China into the following four principal categories, which follow China's classification terminology.

(1) Indigenous coke ovens;
(2) Modified-indigenous coke ovens;
(3) Machinery by-product coke ovens; and
(4) "Clean" coke ovens, earlier called "nonrecovery" coke ovens.

Table 3.2: Major Coal Types in China and Caking Property

Coal Types		Caking Property
Lignite		Has no caking property, hence it cannot be used to make coke by itself. A small amount of ligneous coal can be blended into coking coal and rich coal during cokemaking.
Bituminous Coal	Gas coal	Coke made by gas coal has a small size and slender shape, as well as many longitudinal cracks; besides, the strength is low. Gas coal can be used as blending coal in order to increase shrinkage, reduce the expansion stress, and improve the production ratio of chemicals and coal gas.
	Rich coal	Good caking property. Coke made only by rich coal is easy to melt, but it has a lot of beehive holes and transverse cracks; thus, it is easy to crack. If rich coal makes up a large proportion of the coal used in the coal blending, some adjusters or noncoking coal should be added in order to improve the coke quality.
	Coking coal	Can be used to make coke by itself. Coke made only by coking coal has a good caking property: less cracking, high strength, good abrasiveness, and even size, but it has a high expansion stress.
	Lean coal	Has a weak caking property or even does not cake at all. Coke made only by lean coal is large in size, but is weak in abrasiveness. Hence, lean coal can only be used as an adjuster in cokemaking.
	Open-burning coal	Cannot be used to make coke by itself. If it is blended into coking coal, coke's abrasiveness will be decreased.
	Meager coal	Does not cake, can be used as a contraction agent in coke making.
Anthracite		Does not cake.

Source: Li, Zhehao (1996).

In the United States, the second, third and fourth are called beehive, slot ovens, and nonrecovery, respectively.

In Table 3.3, we provide output data only for three categories, combining output from indigenous and modified indigenous ovens into a single category and calling it modified indigenous ovens, because authorities stopped listing the former designation separately in the statistical publications in the most recent years.

Although oven designs in each category vary, they share general technical characteristics (summarized below).

3.1.1 Indigenous Cokemaking Ovens

"Pile" and "kiln" methods are the oldest and most simple indigenous methods of making coke. In the pile method, coke is made from a pile of coal that is ignited at the bottom. Because there is no coke-oven body, the loss of heat is significant, the process of heating is slow, and large quantities of pollution are emitted to the air, land, and water.

Because there is an oven body to contain the heat and reduce pollution, cokemaking using a simple kiln oven wastes less heat and produces less pollution than cokemaking using a pile. The body of the kiln, however, is simply constructed, and there is no

Table 3.3: Shanxi Province Coke Production and Mix

Unit: million tonnes

Year	China[a]	Shanxi[b]	Machinery[b]	"Clean Coke Ovens"[c]	Indigenous and Modified-Indigenous: Quantity (tonnes)	Indigenous and Modified-Indigenous: Percent of total (%)
1990	73.28	16.09	4.94	n/a	11.15	69.3
1991	73.52	14.55	4.78	n/a	9.77	67.1
1992	79.84	18.15	4.54	n/a	13.61	75.0
1993	93.20	27.51	5.55	n/a	21.96	79.8
1994	114.77	42.79	6.45	n/a	36.34	84.9
1995	135.01	52.98	6.54	n/a	46.44	87.7
1996	136.50	53.96	7.48	n/a	46.48	86.1
1997	137.31	52.79	7.84	n/a	44.95	85.1
1998	128.06	57.03	13.31	n/a	43.72	76.7
1999	120.74	49.60	15.74	0.05	33.81	68.2
2000	121.84	49.67	19.52	0.05	30.10	60.6
2001	131.31	49.88	23.46	0.15	26.27	52.7
2002	142.80	58.52	32.36	0.38	25.78	44.1
2003	177.76	67.47	45.63	1.85	19.99	29.6

Source:
a. China Statistics Yearbook, 1991-2004
b. Shanxi Province Statistics Yearbook, 1991-2004
c. Survey results conducted by Shanxi TVE Bureau in October 2003.
Note: The coke production for Indigenous and Modified-Indigenous coke ovens in Shanxi for 1990-1998 is estimated by subtracting the production of machinery coke ovens from the total coke production. For 1999 to 2003, it is calculated by subtracting the production of machinery coke ovens and clean coke ovens from the total coke production. We assume that only TVEs have clean coke ovens.

sealing used to keep the pollution from escaping. (Eisenhut et al. 1991; Photo 5) For this primitive type of cokemaking, coal consumption is usually in the range of 2.0-2.5 tonnes of coal equivalent (tce) per tonne of coke. All coal volatiles are burned or lost, and the quality of the coke produced is low. (World Bank 1994) A coking cycle using this technology typically takes 12 to 15 days (Zhang 1991). In 1990, about 4.6 million tonnes of coke (about one-quarter of TVE coke production in Shanxi Province) were made in these indigenous ovens (Table 3.4); however, by 2005, almost no coke in China is made in such ovens.

Although these two indigenous cokemaking methods (pile and kiln) since 1996 are illegal in China, our 1998 cokemaking survey indicated that almost 10 percent of the TVE cokemaking plants in Shanxi Province were still using indigenous technologies (MRP 1999b). By 2005, most of the indigenous ovens in Shanxi Province are closed, although Xinhua News posted an article in October 2003 showing the pollution from indigenous ovens still in use in the Luliang Mountain area in Shanxi Province (Xinhua News, 2003 21:58, accessed December 2003, http://www.sina.com.cn).

Table 3.4: National TVE Coke Output by Different Types, 1990

Coke Oven Type	Coke Output (million tonnes)	Coke Output (percent)
Primitive Indigenous	4.6	24.7
PX	11.8	63.4
JX-1	1.0	5.4
Types 75, 89	0.7	3.8
Small Mechanical	0.5	2.7
Large Mechanical	0	0
Total	18.6	100.0

Source: World Bank (1994).
Note: PX = Pingxiang; JX-1 = Jiexiu-1; Type 75 = Taiyuan Jixiehua 1975; Type 89 = designed by Sansheng in 1989.

3.1.2 Modified-Indigenous Cokemaking Ovens

Since the 1970s, the Chinese have continuously made improved versions of coke ovens. They called the first improved versions of non-machinery coke ovens "modified-indigenous" ovens (Photos 6-8). Among these modified-indigenous ovens, the earliest were simple brick ovens, called PX (Pingxiang) and LL (Luliang). These ovens have coking chambers that are separated from each other, and they directly discharge into the atmosphere the combusted coke-oven gas (COG).

When coke is made in modified-indigenous ovens with good coking coal (good caking properties and low sulfur and ash content), producers usually obtain high-quality coke (with high levels of fixed carbon, high strength, and low sulfur and ash content), partially because they can tamp the coal in the oven. Even so, the modified-indigenous ovens still produce serious air pollution, because the combustion of volatile matter is incomplete and because the combustion products are usually released about two meters above the ground level. In addition, because of the simple construction of these ovens, the water used for coke quenching may permeate through the top soil and contaminate the groundwater.

During the 1980s, a series of second- and third-generation, modified-indigenous ovens gained popularity in Shanxi Province. Most were similar to the beehive ovens widely used in industrialized countries at the beginning of the 20th century. Ovens like JX-1 (Jiexiu-1) feature simple tar-recovery mechanisms and COG recycling for lighting coke ovens, which for non-machinery coke ovens is the process used to raise the temperature of the oven to create continuous combustion of COG. The ovens are built in long rows, called batteries, with retaining walls between each oven. All ovens in one battery share a high (greater than 40-meter) stack for discharging combusted gas (Li 1995).

We give an example in Figure 3.1 of the JKH-89 ovens, which were designed by Jiexiu Erji Gongsi in the year 1989, with I and II designating the particular model of oven. The JKH-89 designers increased the combustion zone on top of the coal bed in size over that of the JX-1 oven, in order to improve the thermal efficiency of the oven and to realize the complete combustion of gases. Workers seal the top of the JKH-89 oven with a mobile cover, which they slide into place on rails (Figure 3.1). They load the coal through the opening of the mobile cover. After the coke is produced, workers use water to quench it with the cover in place, thus substantially reducing the release of particulates and vapor. The JKH-89 oven has a tar ditch at the bottom, through which it is possible to recover about 12 kilograms (kg) of tar per tonne coke (compared with 5 kg for the PX type). Therefore, tar permeation into the groundwater is effectively reduced from that of the PX oven. The coal consumption of JKH-89 coke ovens is typically under 1.4 tonnes of coal equivalent (tce)/tonne of coke, which is significantly lower than that of indigenous cokemaking (2.5 tce/tonne of coke), but it is still higher than that of large-machinery coke ovens (1.2 tce/tonne of coke) (Table 3.5). The quality of coke produced in this type of oven is generally first- or top-grade metallurgical coke (World Bank 1994).

Table 3.5: Typical Coal Consumption for TVE Coke Ovens, 1990

Coke-Oven Type	Coal Consumption (tce per tonne of coke)
Primitive Indigenous	2.50
PX	1.56
JX-1	1.43
Types 75,89	1.39
Small-Machinery	1.40
Large-Machinery	1.20
Industry average	1.90

Source: World Bank (1994).
Note: PX = Pingxiang; JX-1 = Jiexiu-1; Type 75 = Taiyuan Jixiehua 1975; Type 89 = designed by Sansheng in 1989; tce = tonnes of coal equivalent.

Advanced modified-indigenous ovens, like the JKH-97 (Photo 9), designed in 1997, that Chen (2000) covers in one of his case studies, represented the best non-machinery cokemaking technology available in China when he did the study in 2000. The structure of the JKH-97 oven is similar to that of the JKH-89 oven, but JKH-97 ovens have a better combustion and heat-exchange system than the earlier modified-indigenous coke ovens. Airflow in the oven is easily controlled to reduce the burning of coal. COG is reused for lighting other ovens, and the gas discharged is much cleaner than in the earlier JKH oven.

3.1.3 Machinery By-Product Cokemaking Ovens

By-product, or so-called "recovery-type," coke ovens are designed to collect various valuable substances (e.g., ammonia, benzene, and tar) released from the volatile materials in coal during the coking process. Because these machinery by-product ovens typically have tall and narrow structures, they sometimes are referred to as slot ovens (Table 3.6; Photos 10-11).

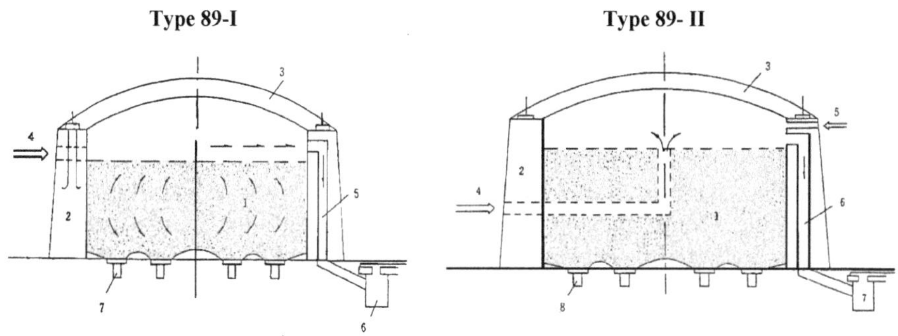

1. Oven Body
2. Oven Wall
3. Mobile Cover
4. Oven-Lighting Hole
5. Flue-Gas Branch Channel
6. Flue-Gas Main Channel
7. Tar Channels

1. Oven Body
2. Oven Wall
3. Mobile Cover
4. Oven-Lighting Hole
5. Secondary Air Intake
6. Flue-Gas Branch Channel
7. Flue-Gas Main Channel
8. Tar Channels

Source: Li, Shaojing and Shen Weiqing (1995: 94, 105).
Note: Both of the JKH Type 89 ovens were designed by Jiexiu Erji Gongsi (second company) in the year 1989, with I and II designating the particular model of oven. These ovens are used by the Sansheng coke company in Jiexiu City as well as other plants in Shanxi Province.

Figure 3.1: JKH Type 89-I and 89-II Coke Ovens

Table 3.6: Typical Pollutants from Cokemaking

Unit: Kilograms per tonne of coke

Pollutants	Indigenous Technology	Modern Technology
NOx	1.4	0.3
SOx	0.2 – 6.5	0.1
Particulate Matter	0.7 – 7.4	0.2
Ammonia	0.1	0.5
Benzene	2.0	0.6

Source: World Bank (1995).
Note: NOx = nitrogen oxide, SOx = sulfur oxide.

Although there are many technical variations, all machinery coke ovens essentially have the same three main parts: (1) coking chambers, (2) heating flues and regenerative chambers—all constructed of refractory bricks, and (3) a steel frame (*China Metallurgical Encyclopedia* 1992). The entire structure is supported either by the ground or by columns under a reinforced concrete or structural steel base. By using the newest machinery coke ovens, coke-plant operators are able to recover most chemical by-products and COG, but for the sake of the environment, they need to be careful concerning the coke oven's maintenance and operation in order to reduce leaks of pollutants during the normal charging and pushing operations (MRP 1999a). Coke-oven doors are especially prone to become deformed and leak.

Partially due to the chamber structure, most coke ovens are not single isolated units. Instead, they are connected to each other to form an organizational unit, the battery. Batteries of machinery coke ovens contain ten to over 100 ovens. In China, batteries of 30-40 ovens are common (*China Metallurgical Encyclopedia* 1992), while in the United States, large batteries of 45 or more ovens are more commonly found. Coking chambers in a battery are separated from and alternate with heating chambers, so that there is a heating chamber on each side of a coking chamber to enhance heat exchange. The regenerative (combustion) chambers are underneath the heating and coking chambers. Separating walls between regenerators also serve as foundation walls for the heating and coking chambers (AISE 1999). Modern by-product coke ovens also have many accessory facilities, such as charging machines, pushing machines, quenching cars, quenching towers, and cooling wharfs. Coke ovens are connected to chimneys through flues, with about 40 ovens per chimney.

The dimensions of a coke oven determine the capacity of the oven, the volume of the "coke-push," and coal consumption. In 1992, the then largest coke oven in the world, which was at Krupp Mannesmann Steelworks in Germany, had an effective volume of 70 cubic meters (m^3) (height = 7.85m, length = 18m, width = 0.55m) and a battery capable of producing 100,000 to 1 million tonnes of coke annually, which is about four times the capacity of the largest battery then operating in Shanxi Province (2nd International Cokemaking Congress 1992). As we noted in Chapter 2 (Table 2.3), by 1995 2% of the TVE plants produced more than 200,000 tonnes annually and by 2000, 12% did, but none of the Shanxi Province TVE plants produced more than 1 million tonnes until recently. In terms of the coal consumption, according to 1994 measurements, small-machinery coke ovens typically consume 1.4 tce/tonne coke (World Bank 1994), but consumption has been improved to around 1.25 tce/tonne coke thereafter (Shen 1999).

In China, during the early 2000s, most of the machinery coke ovens are operated by state-owned enterprises (SOEs), many of which are plants subsidiary to iron- and steel-making enterprises. At present (2005), some township and village enterprises (TVEs) in Shanxi Province are installing machinery coke ovens. However, most of the TVE machinery coke ovens are relatively small (about 50,000-200,000 tonnes coke/year) and the coke quality from these small-machinery coke ovens is usually poorer than that of the modified-indigenous ovens, such as the JKH-89 type, partially because modified-indigenous technology includes a tamping procedure done to the coal bed. According to the 2003 survey of TVE cokemaking in Shanxi Province (UNIDO 2004, p. 3), we find that traditional machinery (slot) TVE coke ovens accounted for 67% of the total output of 33.4 million tonnes, that modified indigenous coke ovens (representing 69% of the total coke plants) accounted for only 28% of the total coke output, and the so-called "clean" ovens (discussed next) accounted for only 5%.

3.1.4 Clean (Nonrecovery) Cokemaking Ovens

As for the newly developed clean type of coke ovens, some confusion exists in the terminology. Analysts refer to them as "nonrecovery," "heat-recovery," or "clean" coke ovens. In one sense, the earliest indigenous, and even modified indigenous, ovens were nonrecovery in that all, or most, of the COG, tars, and other pollutants were emitted to the air, land, and/or water, thus not recovered.

The term "nonrecovery" was originally used by the Sun Coke Company in the United States for a particular type of low-pollution coke oven. In 1960, they built three test ovens at a plant at Vansant, Virginia, and by 1962 they began using Jewell-Thompson coke ovens. Sun Coke has built eight generations of these, negative-pressure, heat-recovery ovens (HROs). The negative pressure pulls air into the oven and helps keep the toxic chemicals from escaping. In 1999, Sun Coke set up a new plant at Indiana Harbor, East Chicago, Indiana, with the two plants combined producing about 2 million short tons annually. (Barkdoll and Westbrook 2001) In March 2005, they began operation at a 550,000 tons-per-year facility in Haverhill, Ohio (Sun Coke, accessed 05/23/05 http://www.sunocoinc.com/aboutsunoco/suncoke.htm).

When we first visited Shanxi Province in 1998, the local Environmental Protection Bureau officials were aware of these HROs developed by Sun Coke in the United States and wanted more information about them, partially because they knew that in 1990, the U.S. Environmental Agency gave the Sun Coke HRO the MACT (Maximum Achievable Clean Technology) designation (Barkdoll and Westbrook, 2001: 1-3).

The TJ-75 (Taiyuan Jixiehua 1975) is the first non-machinery, nonrecovery coke oven successfully used in China. The TJ-75 modified-series ovens are horizontal, cave-dwelling style, high-temperature ovens (Figure 3.2). The ovens have arched tops and double chambers that abut each other, reducing the need for the extra divider wall required in single-chamber ovens. The operator ignites the coke oven through several lighting holes, and COG is gathered in the space between the oven ceiling and the top of the coal pile. Secondary-air intakes are designed to encourage efficient combustion. Mixed with secondary air, COG flows along the combustion channels in the wall and bottom of the coke oven, continues the combustion, and heats the coke oven. Flue gas can be emitted into the air, or an operator can redirect it by use of a valve. There is almost no steel material used in the early versions of this type of oven, and no electrical equipment is needed. (Chinese Department of Agriculture 1991)

The TJ-75 coke ovens feature better operational stability than other modified ovens and consume 0.5 tonnes of coal per tonne of coke produced less than the indigenous oven. The ovens' shortcomings are excessive space above the coal within the oven chamber that causes low thermal efficiency and a slow in-oven temperature rise because of its one-sided ignition system (World Bank 1994).

The SJ-96 (Sanjia 1996) oven represents another of the earliest nonrecovery coke ovens in China (Photos 14-15). In 2000, Sanjia Coke Plant had both a nonheat recovery plant in Jiexiu and a heat-recovery SJ-96 plant was being put into operation with 144 ovens at their Ruiyude plant, which is also at Jiexiu. By 2004, the managers had installed a heat-recovery system in Jiexiu as well (Photo 16).

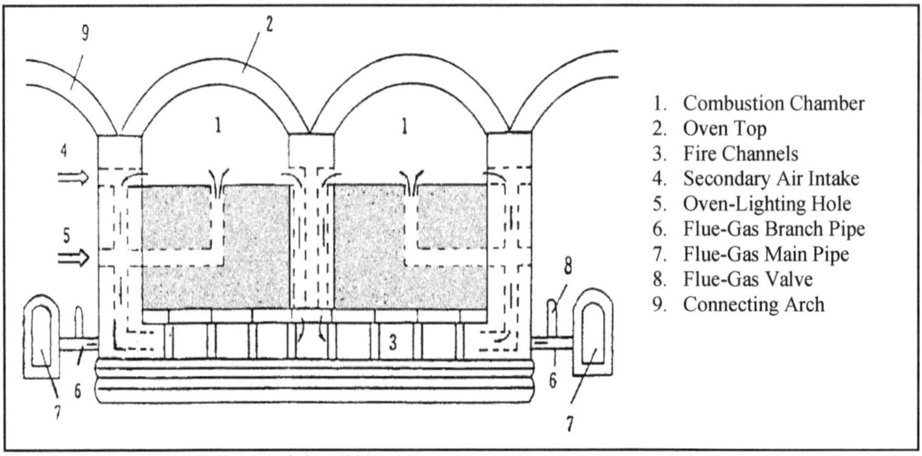

Source: Li, Shaojing and Shen Weiqing (1995: 91).
Note: Type 75 (Taiyuan Jixiehua 1975); thus, it is also known as TJ-75.

Figure 3.2: Taiyuan Jixiehua Type 75 (TJ-75) Coke Oven

In the 1990s, Shanxi Province officials used the term nonrecovery for these ovens, but they soon switched to calling the ovens "clean coke ovens," partly because the term nonrecovery was a misnomer. From the newest of these clean ovens, the plant recovers heat to use for steam and electric-power generation. The QRD-2000 (Qingjie Rehuishou Daogu Lu--clean oven with tamping and heat recovery) and QRD-2002 are both clean coke ovens with heat recovery, while the DQJ-50 (Daogu Qingjie Lu--tamping clean oven, with 50 tonnes capacity per battery) has no heat recovery. Early versions of the SJ-96 (Sanjia 1996; Photo 15) did not have heat recovery, but as of 2003, the SJ-96 ovens at Sanjia have been redesigned and now they have a heat-recovery system (MRP 2004).

The number of clean coke ovens in Shanxi Province is relatively small. As of October 2003, of the 589 coke plants we surveyed in Shanxi Province, 8 plants had clean coke ovens and 174 machinery coke ovens, while 407 had modified-indigenous coke ovens (UNIDO 2004). The current rationale of the local regional officials in China for promoting clean coke ovens is that reducing pollution has priority, and these coke ovens are environmentally superior to the chemical by-product coke ovens that are still prevalent throughout the region. In addition, the by-products currently have low prices, so that the plant managers do not see any benefit in having the by-products recovered. Although the pollution is minimal from clean coke ovens, we note that all the tar and chemicals, such as ammonia, benzene, etc., are fully combusted along with all the toxic pollutants, and those products can never be recovered, raising a potential problem in the future as the need for such chemicals increases. (MRP 2004)

By 2005, as indicated in Chapter 2, there are a number of nonrecovery coke ovens, almost all of which have heat-recovery units; in Shanxi Province, these are called "clean coke ovens" rather than "nonrecovery," because they do recover heat. Plants with these ovens burn all volatile materials and do not collect any by-products. We have visited both the HRO Sun Coke plant at Indiana Harbor in the United States and a clean coke plant in Houma, Shanxi Province, China, where there is a similar type of heat-recovery oven. A visual difference is that there was considerable observable pollution escaping from the air holes at the Houma plant and none from the Indiana Harbor plant, although both plants were about the same age when visited. The Houma plant manager was aware of the problems and they were constructing new ovens that were not yet in operation. (MRP 1998b; MRP 2004; Photo 17) We have also visited two of the latest clean coke ovens in Shanxi Province (Photos 18-20). By 2005, many more clean coke ovens are being constructed in Shanxi Province, most of these designed by Chinese coke-oven design institutes. One design institute alone, for example, already has 20 batteries in operation and another 10 under construction (Fang 2005).

3.2 Internal Supply Chain of Alternative Cokemaking Technologies

At the industry level, the coke-making processes include a series of stages along the supply chain from coal mining to coke utilization (Figure 3.3), including:

- Shipping,
- Storing,
- Coal preparation,
- Coal charging,
- Coking,
- Coke pushing,
- Coke quenching,
- Coke crushing and screening; and
- Coke shipping to final users.

At each stage along the cokemaking supply chain, the coke producer uses input materials and labor and/or transforms them into products and wastes (Polenske 2000). Within a cokemaking plant, the internal supply chain includes coal preparation, coking (in coke ovens), by-product processing, and coke screening. By focusing on the internal supply chain, we can compare the technical differences between the aforementioned cokemaking technologies.

3.2.1 Coal Preparation

Raw coal taken directly from the coalmines generally has high ash and sulfur content, which is unsuitable for many applications, including cokemaking. At the Dongshan Coal Mine (MRP 2000, July 7; Photo 25), for example, we were told that they reduced the ash content form 25% to 15% by washing the coal. In Shanxi Province, the ash content of coking coal ranges from 15% to 18% and the sulfur content ranges from 1% to 2%, both of which make the coking coal in Shanxi Province of much higher quality than in most other countries. For example, in India, the ash content is as high as 25% and the sulfur content as high as 3%. (MRP 1999a)

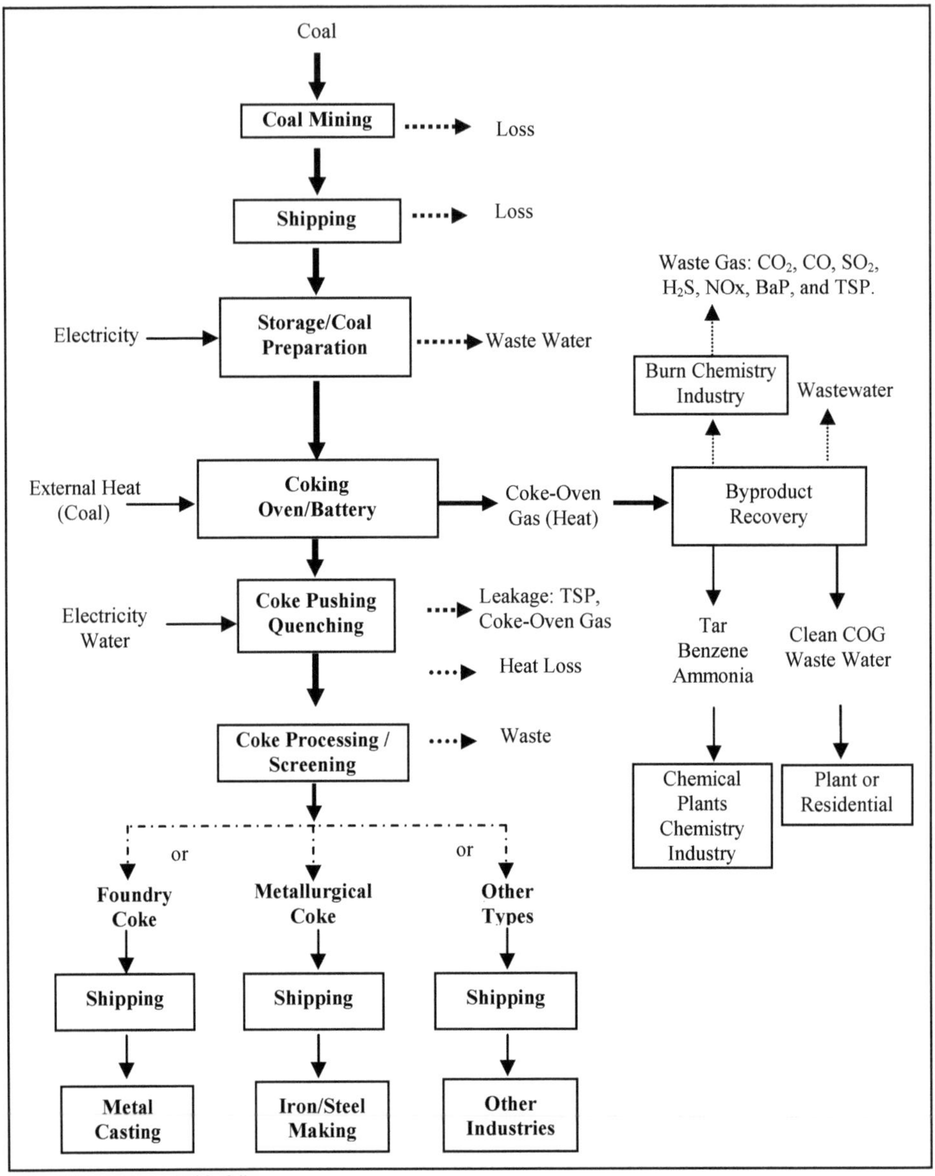

Source: Chen, Hao (2000: 24).
Note: CO_2=carbon dioxide, CO=carbon monoxide, SO_2=sulfur dioxide, H_2S=hydrogen sulfide, NOx= nitrogen oxide, BaP=Benzopyrene-3.4, TSP=total suspended particulates.

Figure 3.3: Flowchart of the Internal Cokemaking Supply Chain

Before coal can be used for cokemaking, coke producers usually crush, clean (or wash), and blend it. These processes help the coke producer reduce coal to an appropriate size, reduce ash and sulfur content in coal, and conserve high-quality coking coal by mixing up to four types of coal during the blending procedure.

By coal cleaning, the first major step of coal preparation, the worker effectively removes inert materials (like ash) and undesirable material (like nonorganically bound sulfur) from the coal. In Shanxi Province, most raw coal is obtained from nearby mines that are 25-100 kilometers from the cokemaking plants and is usually shipped to the plant by truck and cleaned in the plant. In other provinces in China, cokemaking plants may buy pre-cleaned coal, which reduces shipping costs on the non-productive materials, such as dirt, ash, and sulfur, but in Shanxi Province, because water is scarce near the mines, coal is rarely cleaned at the mine. Although coal mined in the southern part of Shanxi Province is considered to be good coking coal (because its ash and sulfur content is relatively low), the coal still needs to be cleaned in order to make high-quality coke. (MRP 1999a)

In the coal-cleaning stage, workers can use physical, chemical, or biological operations that upgrade the quality of coal by regulating the size and reducing ash, sulfur, and other impurities. China would be able to use one or more of the four following types of coal-cleaning methods:

1. Conventional physical methods: mechanical operations and a heavy medium are used to separate coal with different ash content;

2. Advanced physical methods: electromagnetic or other methods are used to separate coal;

3. Chemical coal-cleaning methods: chemicals are used to reduce undesirable contents in the coal; and

4. Biological methods: biological reactions are used to remove sulfur from the coal.

In China, conventional physical coal cleaning is the most widely used method, because of its cost effectiveness, but those using the process cannot remove organically bound sulfur (World Bank, accessed on 01-18-05, http://www.worldbank.org/ html/ fpd/ em/power/EA/mitigatn/aqsocc.stm). Usually, physical methods involve manual sorting, selective crushing, sizing, washing, dewatering, drying, and classification (Zhang 1991; Photos 29-30). Other coal-cleaning methods, such as advanced froth floatation, electrostatic, and heavy liquid cycloning, require expensive equipment and do not improve the cleaning capacity significantly; thus, they are seldom used in China. Even so, organically bound sulfur can only be removed by chemical or biological methods. In general, coal cleaning can reduce the average ash content of coal from 19 percent to 10 percent and reduce sulfur content from 1.0 percent to 0.7 percent (China Environment Yearbook 1997).

3.2.2 Coking

Cokemaking practices differ from country to country and from plant to plant. In China, they differ especially for indigenous and modified-indigenous cokemaking. Here, we provide a general introduction to the operations and equipment used in most of the cokemaking practices throughout the world, with specific reference to China, where relevant. Most coking practices involve the following steps: charging, leveling, sealing, heating, pushing, and quenching.

3.2.2a Charging

Charging describes the process of feeding prepared (washed and sorted) coal, mostly in small pieces of less than 3 mm (millimeters), into coke ovens (Li 1996). In a modern machinery cokemaking plant, coal is fed into a series of inter-linked ovens (battery) by machines. Larry cars or a loading machine on top of the batteries take coal from storage bins to the coke ovens and fill the ovens through charging holes at the top of the ovens (Eisenhut 1991; Photo 28). For the indigenous and modified-indigenous cokemaking in China, coal is either poured into coke ovens through the uncovered roof or filled through the open end of the oven either by manual labor, or skip-bucket trucks, or a conveyer belt (MRP 1998).

3.2.2b Leveling

Leveling refers to the process of leveling coal within a coke oven to a given height, evenly distributing it throughout the oven. The height of the coke pile should provide a space above the bed of coal to allow for the collection of gases emitted during the coking process. In machinery cokemaking, workers use a pusher machine that extends a long steel ram into the slot oven to level the coal pile. For the indigenous and modified-indigenous ovens in China, laborers level the coal using a rake, or sometimes tamp (tightly press down) the coal either with a machine or by hand. By tamping the coal, the workers can improve the strength of coke, which is good for the applications in iron and steel making. (MRP 1998; MRP 2004)

3.2.2c Sealing

In some indigenous coke ovens in China, there are openings on the top of the ovens for gas discharging. Workers use bricks and mud to seal coke-oven doors in many of these indigenous ovens. For machinery (slot) ovens, the coal is loaded into the oven, and the sealing is done before heating the coal-filled ovens. By sealing the coke oven, the workers provide a means to separate the gas that accumulates inside the coke oven from the outside. The purpose is both to prevent oxygen in the air from getting into the coke oven and to contain COG emissions (Photo 41). For the machinery cokemaking plants, coke-oven doors and covers are made of steel, and workers operate them mechanically. As a machinery coke-oven ages, the seal on these doors and covers can leak gases and particles, which is one of the largest sources of air pollution at a machinery coke plant. (MRP 1998).

3.2.2d Heating Methods

Most indigenous and modified-indigenous coke ovens in China are heated by the direct combustion of the COG in the coal-carbonization chamber. Heating methods used for machinery coke ovens are completely different from those used for indigenous coke ovens, but the main source of heat is the same--the combustion of COG. In machinery coke ovens, the combustion of COG occurs in the chambers around a coal-carbonization chamber, and the coke oven is heated by heat exchanged through the chamber walls (Eisenhut 1991). Part of the coking coal may be burned during the coking process because of the direct combustion of COG in the oven (MRP 1998).

Coke is made slowly in indigenous and modified-indigenous ovens--up to 10 days is common in China. For the machinery coke ovens, the coking time per oven is usually much shorter--14 to 36 hours, depending on the battery design and the way it is operated. Although the volume of coal charging in modified-indigenous coke ovens (up to 160 m^3) is usually much larger than that of machinery coke ovens (up to 70 m^3), the production rate of machinery cokemaking is still usually more than that of modified-indigenous coke ovens (MRP 1998, 1999a).

3.2.2e Pushing

In most of the modified cokemaking plants and some of the clean coke ovens, workers do not use a coke-pushing operation. Instead, they directly quench the coke in the coke oven and remove it from the oven at a low temperature either manually or with a small tractor with a fork lift (MRP 1998). In machinery cokemaking, when the coal is fully coked, workers open the coke-oven doors with a machine, and they use a pusher machine to push the coke from the oven. They use a large ram, mounted on the pusher machine, to push the coke from the "pusher side" out through the “coke side” of the oven (AISE 1999). A coke guide on the coke side directs the coke across a bench and into a quenching car (Photo 11). The large cake of baked coke breaks when it is cooled or handled, and coke breeze (fine coke particles) is generated. During hot pushing used in machinery cokemaking, workers use the pusher machine to push the coke from the oven while the coke is at a high temperature, which reduces the time needed for pushing coke and charging coal between two consecutive rounds of coke production in one oven.

3.2.2f Quenching

As noted earlier, in cases of Chinese indigenous and modified-indigenous cokemaking, workers do not use a quenching car, because they quench the hot coke with water while the coke is still inside the oven. In most machinery cokemaking plants in China, workers use a quenching car to transfer the hot coke to a nearby quenching tower, where a large amount of water is poured onto the coke to cool it (Photo 13). During the quenching operation, coke cracks because of the cooling of the surface. Although a lot of steam and particles are generated, in most Shanxi Province cokemaking plants, as of 2005, only a few used emission-control devices on the quenching car and quenching tower. In plants in the United States and some other countries, special mobile units, called scrubber cars, are used to capture emissions, or the pushing is done in a shed that has vents to trap the air pollutants emitted. The coke is dumped from the quenching car onto a cooling wharf before being transported usually by a conveyor belt to the storage area (Photo 12). Another advanced quenching method is to circulate an inert gas (such as nitrogen) around hot coke, which is called dry quenching (AISE 1999). This process is just in 2004 being considered for use in Shanxi Province mainly because it is expensive, but it is being used in steel plants in Liaoning Province. (MRP 1998; MRP 2003; MRP 2004)

In summary, these coke-oven processes involve handling bulk material and raising and lowering the temperature of the coal and coke. The operations performed during these processes not only determine the quality of coke produced, but they also have a large impact on the pollution generated by cokemaking. Clean coke ovens have the least air, water, and land pollution; well-maintained machinery coke ovens can meet many of the current environmental regulations for air pollutants; most modified indigenous ovens have serious pollution emissions; and the indigenous ovens are worse still. Because modified-indigenous cokemaking plants use more labor and less machinery than machinery and clean cokemaking plants, they can be more profitable. (MRP 1998, Polenske and McMichael 2003)

3.2.3 By-Product Processing

In indigenous and modified-indigenous ovens in China, most managers have not yet installed by-product facilities; thus, the COG is often discharged directly into the air or discharged after further combustion of the gas (MRP 1998). In modern coke ovens, plant managers use standpipes, or ascension pipes, lined by refractory bricks (a special type of fireproof brick) to carry away the volatiles created by heating the coal during the coking process. The pipes run from each oven to a collecting main. In by-product coke ovens, pipes carry COG to the chemical-processing facilities, where gases and other chemicals are recovered. (AISE 1999)

In some modern coking plants, managers distill the coal to produce coke and fuel gas. When producing 1.0 tonne of coke, about 130 kg of COG are generated (Lupis and McMichael 1999). When COG is cooled, managers can recover some by-products, such as tar, ammonia, and benzene, which are valuable chemicals; however, these by-products can become potential pollutants if they are not collected or handled carefully (Li 1995).

"Off gases" from the cokemaking operations contain impurities including ammonia, hydrogen sulfide, and hydrogen cyanide, all of which must be removed before the gas is distributed as fuel (*China Metallurgical Encyclopedia* 1992). In most large-machinery cokemaking plants in China, flushing liquor formed from water sprayed into COG contains tar. Flushing liquor and liquor formed in primary coolers is sent to a tar decanter for tar recovery and for separating the ammonia liquor. In some facilities, managers use an on-site tar distillation unit to process tar further. Ammonia liquor is separated from the tar decanter, and the remaining wastewater is sent to a wastewater-treatment site after ammonia recovery. Further removal of tar from the COG is accomplished with an electrostatic tar precipitator. COG is further cooled in a final cooler, where naphthalene is removed. Light oil recovered from COG is used as an absorbent to recover benzene, toluene, and xylene (World Bank 1995). Other processes are used for removing sulfur from COG. In the case of modified-indigenous plants in China, managers generally recover only tar and light oil from the liquor formed by cooling the COG (MRP 1999a).

Plant managers can have organic materials, such as tar and naphthalene, produced from by-product recovery, separated, and then sold. Alternatively, they can handle the materials as waste products. For the raw gas (mainly carbon dioxide, carbon monoxide, and water vapor), managers may decide to clean it more and supply it as an industrial or residential fuel, or they may decide it can be combusted and vented to the atmosphere (World Bank 1995). In most modified-indigenous plants in China, only limited, if any, recovery of by-products is attempted (MRP 1998).

3.2.4 Coke Screening

To meet the requirements of coke users, suppliers crush large lumps of coke into smaller ones and then screen them by size. At modified-indigenous cokemaking plants in China, laborers break the coke into smaller pieces using large sledgehammers (MRP 1998), while in the machinery cokemaking plants, machines crush the coke, which workers then put through a series of sifters of different sizes. The lump-size requirement of coke in China is 40 mm for large blast furnaces and 25-40 mm for medium and small furnaces (Chen 1999). In modified-indigenous plants in China, workers pick out the high-quality, appropriate size coke manually to improve the overall quality of the coke produced.

3.3 Conclusion

In this chapter, we have reviewed the four major cokemaking technologies used in Shanxi Province since the 1980s. We emphasized the characteristics of several generations of these coke ovens used in the region. Today (2005), as noted in Chapter 1, Shanxi Province is the largest producer of coke in China, and the demand in China for coke is increasing as automobile, construction, and other industrial production requiring steel rapidly expands. Shanxi Province is also an international producer, selling its coke throughout the world. With the development of new cokemaking technologies, Shanxi cokemaking plant managers are in a continuous process of updating coke ovens and related equipment.

During the past 10 years, we have seen the TVE plants become the dominant producers of coke in Shanxi Province. We will look at these TVE cokemaking plants in more detail in Chapter 4, where we compare the energy efficiency of China and Shanxi Province, and in Chapter 5, where we specifically compare SOEs and TVEs.

References

AISE. 1999. *The Making, Shaping, and Treating of Steel, 11th Edition. Ironmaking Volume*. Pittsburgh, PA: The AISE Steel Foundation.

Barkdoll, Michael P., and Richard W. Westbrook. 2001. Consistent Coke Quality from Sun Coke Company's Heat Recovery Cokemaking Technology. *Coke Summit, 2001*. Portland, ME: Intertech, Session 5.

Chen Hao. 2000. Technological Evaluation and Policy Analysis for Cokemaking: A Case Study of Cokemaking Plants in Shanxi Province, China. Master of Science Thesis. Cambridge, MA: Technology and Policy Program, Massachusetts Institute of Technology (May).

Fang, Jinghua. 2005. E-mail correspondence (April 17).

Li, Shaojing and Shen Weiqing. 1995. *Production and Pollution Control of Modified-Indigenous Coke Ovens.*Taiyuan: Shanxi Science and Technology Press.

Li, Zhehao. 1996. *Coke Making Questions and Answers*. (no publisher listed)

Lupis, Claude H.P. and Francis C. McMichael. 1999. Input-Output Microeconomic Model for Process and Environmental Analysis. *Proceedings, Global Symposium on Recycling, Waste Treatment, and Clean Technology*, San Sebastian, Spain.

MRP (Multiregional Planning) Group, MIT. 2004. Notes: Field trip to Shanxi Province,China (January).

MRP (Multiregional Planning) Group, MIT. 2003. Notes: Field trip to Shanxi Province, China (January).

MRP (Multiregional Planning) Group, MIT. 2002. Notes: Field trip to Shanxi Province, China (July).

MRP (Multiregional Planning) Group, MIT. 2001. Notes: Field trip to Shanxi Province, China (July).

MRP (Multiregional Planning) Group, MIT. 2000. Notes: Field trip to Shanxi Province, China (July).

MRP (Multiregional Planning) Group, MIT. 1999a. Notes: Field trip to Shanxi Province, China (January).

MRP (Multiregional Planning) Group, MIT. 1999b. Notes: Field trip to Shanxi Province, China (August).

MRP (Multiregional Planning) Group, MIT. 1998. Notes: Field trip to Shanxi Province, China (August).

Polenske, Karen R., and Francis C. McMichael. 2002. A Chinese Cokemaking Process-Flow Model for Energy and Environmental Analyses. Energy Policy, **30**(10): 865-883.

Shanxi Province Statistics Bureau. Shanxi Province Statistical Yearbook, 1994-1999. Beijing: China Statistics Press.

World Bank. 1994. World Bank Report No. 168/94. Washington, DC: The World Bank.

World Bank. 1995. Pollution Prevention and Abatement Handbook: Coke Manufacturing Section. Washington, DC: The World Bank.

---

[1] At time of writing, Multiregional Planning (MRP) Research staff, MIT, Cambridge, MA, USA; currently, Developer, IntraLinks, Inc., Boston MA.

[2] Professor of Regional Political Economy and Planning. Head, China Cokemaking Team, Department of Urban Studies and Planning, Massachusetts Institute of Technology, USA.

CHAPTER 4

ENERGY-INTENSITY STRUCTURAL DECOMPOSITION ANALYSIS: CHINA AND SHANXI PROVINCE

Ali SHIRVANI-MAHDAVI,[1]
GUO Wei,[1] and Karen R. POLENSKE[1]

4.0 Background

Analyzing energy intensity in China is an important topic. We define energy intensity as the amount of energy consumed to produce one unit of output, where for purposes of this study, we define output as gross domestic product (GDP). Although both energy consumption and output levels in China have grown rapidly during the past 20 years, energy-intensity has decreased substantially during the same period. Furthermore, we selected Shanxi Province for a comparative analysis for three reasons. First, Shanxi Province produces over 25 percent of China's coal, making it China's largest coal-producing province. Second, energy-intensity levels in Shanxi Province are significantly greater than in China as a whole, and third, despite its status as a large energy consumer and producer, energy-intensity levels in Shanxi Province have also dropped significantly during the past 20 years. As such, by analyzing the underlying factors that contribute to energy-intensity levels in Shanxi Province and China, both temporally and spatially, we can understand some of the main reasons behind the decline in energy-intensity in both places.

In this chapter, we describe two methods we used for the study, namely, Structural Decomposition Analysis (SDA) and Spatial Structural Decomposition Analysis (SSDA). We use them to explain the differences in energy-consumption levels in 28 nonenergy and five energy sectors between Shanxi Province and China in 1995 and the changes within Shanxi Province between 1992 and 1999, respectively, the latest years for which we had their input-output tables at the time of the study. SDA is a practical tool that makes it possible to quantify fundamental factors of change in a wide range of

Karen R. Polenske (ed.), The Technology-Energy-Environmental-Health (TEEH) Chain in China: A Case Study of Cokemaking, 41–70.

variables, including economic growth, energy use, employment, trade, and material intensity of use.

In each of the two regions (China and Shanxi Province), this decrease in energy intensity has occurred from 1986-1995 usually in each of six material-producing sectors, with notable exception of the commerce and construction sectors in Shanxi Province (Figure 4.1). There are four facts that stand out from these charts. First, there was a steady decline in energy intensity in most of the six sectors in both regions during this ten-year span; however, the rate of decline differed significantly among the sectors. The light and heavy industry sectors showed marked declines in energy intensity, while the remaining sectors showed a more moderate rate of decline. At the same time, the light and heavy industrial sectors began this period with much higher energy-intensity levels than the remaining four sectors, and, as a result, they had a great deal more room for improving their energy intensities.

The second fact is the major energy-intensity differences in the transportation sector between Shanxi Province and China. Specifically, the transportation sector is by far the most energy-intensive sector in Shanxi Province, starting with 3,500 grams of standard coal equivalent (gsce) in 1986, but showing a marked decline of just under 50% to about 1800 gsce in 1995. In China, transportation also has a high energy-intensity level of just under 1,000 gsce in 1986, and it decreases by about 50% to 500 gsce in 1995.

The third energy-intensity fact in both China and Shanxi Province is that unlike the other five sectors, the commerce sector shows an increase in energy intensity during this ten-year span. We have not yet conducted research as to why this sector had such a dramatic increase. There is an additional difference between the two, with Shanxi Province showing a much larger increase than China in energy intensity in this sector.

The fourth, and most important, fact is the distinct quantitative energy-intensity differences between China and Shanxi Province. A quick glance at the six charts reveals this profound difference. In 1986, the most energy-intensive industry in China, heavy industry, consumed about 900 gsce per 1980 Renminbi (RMB) of GDP. By comparison, heavy industry in Shanxi Province consumed about 1,400 gsce per 1980 RMB of GDP, and Shanxi Province's most energy-intensive sector, the transportation sector, consumed over 3,500 gsce per 1980 RMB of GDP. That figure is almost 4 times as much as the transportation sector in China as a whole in that year. The primary purpose of this study is precisely to understand the factors behind these differences and changes in energy intensity, both temporally and spatially, using SSA and SSDA.

As stated earlier, we utilize SSDA methods to analyze the underlying factors behind the differences in energy-intensity levels between China and Shanxi Province. The conceptual foundation of SDA is input-output economics. Input-output analysis, first introduced by Leontief (1936), is specifically designed as a tool for systematic analysis of mutual interdependencies between different parts of the economy. The empirical basis of input-output analysis is the transactions table, which provides a detailed statistical account of the flows of goods and services among all the producing and consuming sectors of a given economy, that is, among all the various branches of

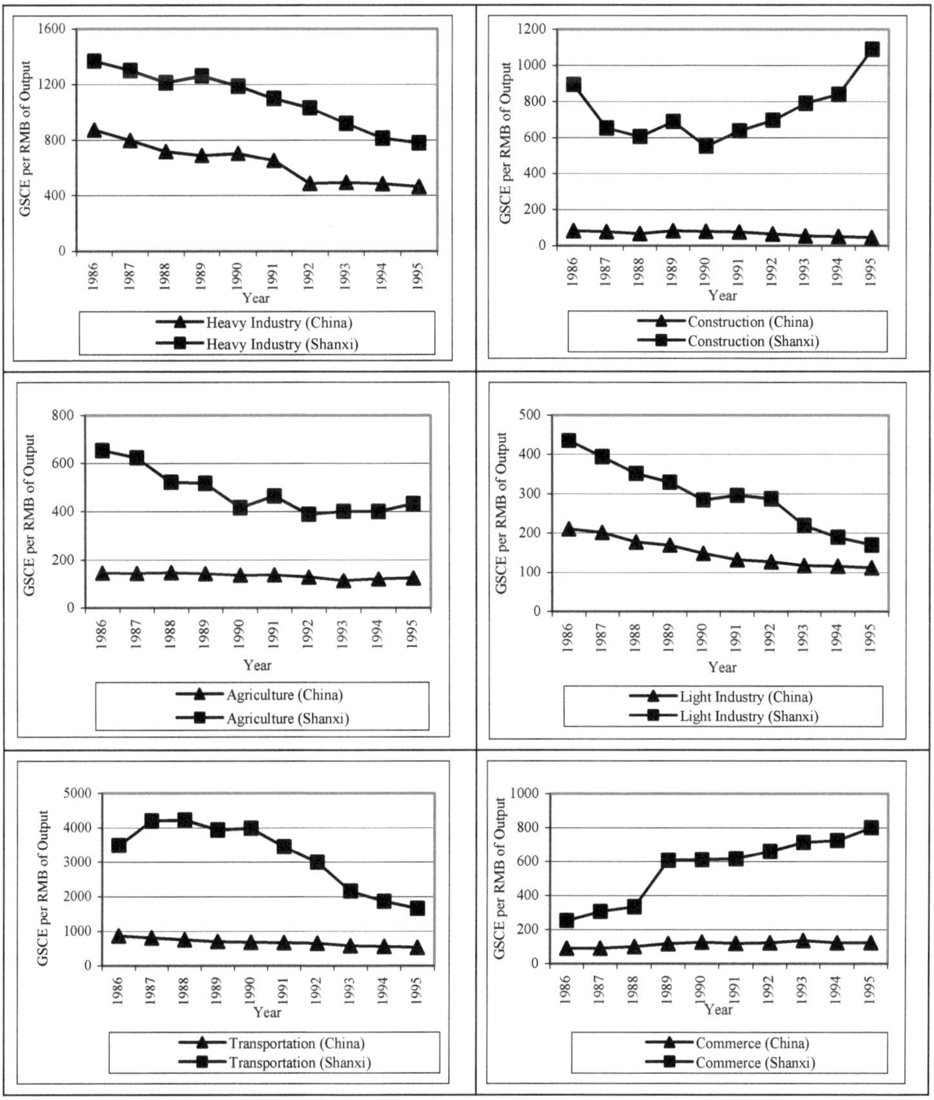

Sources: SSB 1997b. *Gross Domestic Product, 1952-1995:* 192-199; SSB 1991. *China Provincial Statistics, 1949-1989*: 66-67; SSB 1993. *China Energy Statistical Yearbook, 1991*: 205-206; SSB 1996. *China Energy Annual Review, 1996*: 95; SSB 1994. *China Energy Annual Review, 1994*: 131; SSB 1997. *China Statistical Yearbook, 1997*: 216; Shanxi Province 1997. *Shanxi Province Statistical Yearbook, 1997*: 120.
Unit: GSCE per 1980 RMB of output.
Note: RMB: 1 Renminbi, the Chinese currency = 0.125 U.S. Dollar; GSCE: Grams of Standard Coal Equivalent.

Figure 4.1: Energy Intensity for China and Shanxi Province's Material Production Sectors, 1986-1995

business, households, and government. The table displays not only complete details of the income and product accounts, but also all intermediate transactions among producers and purchasers within a consistent accounting framework.

The basic rationale for SDA is to split an identity into its components. This division can be as simple as a three-part basic form or as complex as desired. According to Rose and Casler (1996: 44) "the advantage of this formal derivation is that it ensures that equations have the desirable properties of being (1) mutually exclusive and (2) completely exhaustive."

SDA is a popular tool of analysis for several reasons. First, it overcomes many of the static features of input-output models and can be used to examine changes over time in technical coefficients and sectoral mix. Thus far, we believe that it has only been used for historical and current analyses, but Rose and Casler (1996) indicate how it might be used as a forecasting tool.

Second, SDA is a practical alternative to econometric estimation. Analyses of similar topics using econometrics require a time-series covering fifteen years or more, and not only for output and primary factors of production but for all intermediate inputs as well. On the other hand, SDA requires only two input-output tables: one for the initial year and one for the terminal year of analysis.

Still a third reason for using SDA arises from its input-output base, which is its comprehensive accounting of all inputs in production. This is especially appropriate in the environmental arena, where, as environmental and natural-resource issues become more prominent and serious, there is a greater need to look at the root causes of pollution and resource depletion. In SDA, these causes are more readily linked to intermediate sectors that are ignored in the more standard economic approaches.

Energy is one of the areas where different types of SDA have been used extensively. Strout (1966) analyzed how changes in technology and in the level and composition of final demand affected the United States (US) energy-use changes from 1939 to 1954. Reardon (1976) conducted an input-output analysis of US energy-use changes from 1947 to 1958, 1958 to 1963, and 1963 to 1967. Hannon (1983) compared the energy costs of providing goods and services in the United States in 1963 and 1980. Proops (1984) decomposed changes in the energy-output ratio into three factors: changes in energy intensities, changes in final demand, and changes in the structure of interindustry flows. Ploger (1985) assessed the effects of changes in output mix and energy coefficients on energy consumption in the Danish manufacturing industries. The Office of Technology Assessment (OTA) (1991) staff performed an SDA on US energy-use changes between 1972 and 1988. Finally, and relevant to this study, Lin (1996), and Lin and Polenske (1995) used SDA to explore factors behind the drop in China's energy intensity between 1981 and 1987.

In this chapter, we extend the usage of SDA energy analyses to uncover the factors behind the differences in energy-intensity levels in 28 nonenergy and five energy sectors between Shanxi Province and China in 1995 and the changes within Shanxi Province between 1992 and 1999.

4.1 Methodology

The concept of input-output economics is similar to that of a machine. At one end, energy and other inputs of productive sources are fed into the machine, where they are combined, processed, and transformed. From the other end, a flow of final goods and services emerges. As such, machines with different technologies may require different quantities and combinations of energy and other inputs to produce the same set of final goods and services. The amount of energy required in the production system, therefore, will change if there is a technological change, either due to a modification of the existing production machine, or due to a complete replacement of the old machine with a new one (Lin 1996). The primary factors behind improvements in energy efficiency in China have been and will continue to be the economies of scale, improvements in equipment and processes, and improvements in the fuel mix and industrial raw-material inputs. We note, however, that within the framework of SDA, we are unable to separate the contributions of the individual factors mentioned above.

In this study, we use the direct-input (technical) coefficients in the input-output accounts to describe production technology. We obtain the technical coefficients by dividing the entries in a column of the input-output transaction table, which represents the sector's purchases, by that sector's output. Each coefficient portrays the amount of each particular input required by a given sector to produce one unit of that sector's output. These direct-input coefficients, then, give a quantitative description of the technique of production used by a sector (Carter 1970; Polenske and Fournier 1993). As such, we obtain a systematic tabulation of technologies of all sectors of an economy that provides a concise and detailed description of the technological structure of the economy at a given time (Leontief 1985). We stress, however, that the technology inputs show the average, not the marginal, technology of a sector. Furthermore, within a given sector, there may be a number of heterogeneous subsectors, producing different products and having different input structures. In general, direct inputs show the average input structure of different subsectors, different technologies, and/or different products. We can reduce the heterogeneity of subsectors or products through the use of a more detailed industrial classification. As long as we use the sector as an aggregate, however, there will always be supposed "technology" changes that are actually due to changes in the product mix and/or the changes in the output mix (Carter 1970). In addition, because we use inputs measured in value, rather than physical units, some supposed technology changes will actually be a result of price changes.

The input-output model we use in this study is an open model. The main reason for choosing an open, rather than a partially closed, input-output model is that in China's economy, we cannot assume that the direct relationship between labor income and household consumption and between capital depreciation and gross capital formation is linear and has fixed proportions. In the open model, primary inputs, such as labor and capital (profits, depreciation, etc.) are not included in the calculation of intermediate energy-input requirements. Instead, the energy consumption of workers and capital-investment projects is included as part of final demand. This decision not to include primary inputs in the endogenous portion of the production technology results in an underestimation of the energy embodied in products. The magnitude of the underestimation, however, should be small because about 85 percent of the energy is used as intermediate inputs in China in the 1990s (Lin 1996).

We use SDA to show whether the final energy-consumption in the economy changes because of changes in production technologies or variations in final demand. With the input-output model, we separate energy consumption into two parts: intermediate and final. Intermediate energy consumption is the energy used directly by production sectors as an input. Final energy consumption is the energy used by, or sold directly to, final private and public consumers, such as peasant and nonpeasant households and government agencies. We refer to these as intermediate and final energy consumption, which are represented in Equation 4.1.

$$\mathbf{E} = \mathbf{E_g} + \mathbf{E_d} = e\,[(I-A)^{-1} - I]\,Y + eYn = FY + eYn \tag{4.1}$$

Where $F = e\,[(I-A)^{-1} - I]$, $\mathbf{E_g}$ is the vector of intermediate energy consumption, and Ed is the vector of direct final energy consumption. Equation (4-1) shows that total intermediate-energy requirements, ***F***, and final demand, ***Y***, determine total energy consumption in an economy. ***F***, in turn, is a function of production technology, measured in terms of the technical-coefficient matrix, **A**, which includes both energy and nonenergy inputs.

As such, energy use in the economy can change because of changes in production technology and/or in final demand. Applying Equation (4.1), we describe the changes in energy consumption as:

$$\begin{aligned} \Delta E &= E_2 - E_1 \\ &= (F_2\,Y_2 + e\,Y_2\,n) - (F_1 Y_1 + eY_1\,n) \\ &= (F_2\,Y_2 - F_1\,Y_1) + e\,(Y_2 - Y_1)\,n \end{aligned} \tag{4.2}$$

The first term in Equation (4.2), $(F_2\,Y_2 - F_1\,Y_1)$, represents changes in intermediate and final energy use, which depend on both changes in production technology, F, and changes in final demand, Y. The second term, $e\,(Y_2 - Y_1)\,n$, measures changes in final energy consumption, which is solely a function of final-demand shifts, with the subscript notations, 1 and 2, represent the starting and final points, respectively, of the analysis.

To determine how much of the energy change is due to (1) changes in what to consume (final-demand shifts) or to (2) changes in how to produce (production-technology changes), we introduce a hypothetical economy with starting production technology, F_1, and finishing final demand, Y_2. The energy consumption in this hypothetical economy would be:

$$E_{F1\,Y2} = F_1\,Y_2 + e\,Y_2\,n \tag{4.3}$$

Where $E_{F1\,Y2}$ measures, for instance, the amount of energy that would be consumed in Shanxi Province's economy if the 1992 production technology were used to deliver 1999 final demand. Using $E_{F1\,Y2}$ as a reference point, we rewrite the energy-use change as:

$$
\begin{aligned}
\Delta E &= E_2 + E_{F1\,Y2} - E_{F1\,Y2} - E_1 \\
&= (F_2\,Y_2 + e\,Y_2\,n) + (F_1 Y_2 + e\,Y_2\,n) - (F_1 Y_2 + e\,Y_2\,n) - (F_1 Y_1 + e\,Y_1\,n) \\
&= F_1\,(Y_2 - Y_1) + e\,(Y_2 - Y_1)\,n \qquad \textit{(final-demand shift)} \\
&+ (F_2 - F_1)\,Y_2 \qquad \textit{(production-technology change)} \qquad (4.4)
\end{aligned}
$$

The final-demand shift indicates the energy impact of final-demand changes while production technology is held constant. There are three components to the changes in final demand: (1) level, (2) distribution, and (3) pattern of final demand. The level of final demand refers to the overall level of total final demand, which, in our example, equals the sum of expenditures by final consumers, or the gross domestic product (GDP). The distribution of final demand refers to the allocation of total final demand among the individual final-demand sectors, such as personal consumption, government expenditures, capital investments, exports, and imports. The pattern of final demand refers to the mix of goods and services within individual final-demand sectors, such as personal consumption, is distributed among food, housing, transportation, entertainment, and other expenditures.

Production-technology change quantifies the energy effect of changes in production technology with a given final demand. However, it is not the only reference point that an analyst can use to separate energy-use changes into final-demand-shift and production-technology-change components. An alternative is for the analyst to use E_{F2Y1} as a reference point, i.e., the production technology of year 2 and the final demand of year 1, in which case Equation (4.4) becomes:

$$
\begin{aligned}
\Delta E &= E_2 + E_{F2\,Y1} - E_{F2\,Y1} - E_1 \\
&= F_2\,(Y_2 - Y_1) + e\,(Y_2 - Y_1)\,n \qquad \textit{(final-demand shift)} \\
&+ (F_2 - F_1)\,Y_1 \qquad \textit{(production-technology change)} \qquad (4.5)
\end{aligned}
$$

It is obvious that except by pure chance or under some strict mathematical condition, Equations (4.4) and (4.5) will attribute energy-use changes to final-demand shift and production-technology change differently. This ambiguity stems from the problem of indexing, that is, weights from one year usually do not give the same answer as weights from another year, and there is no single correct answer (Strout 1966; Carter 1970). In fact, the two equations are designed to answer different questions, as is pointed out by Strout. In Equation (4.4), we ask, for example, "how much more (or less) energy would have been required in 1999 if the 1992 production technology had been used to satisfy 1999 final demand?" In Equation (4.5), we ask "how much more (or less) energy would have been required in 1992 if the 1999 production technology had been available to deliver 1992 final demand?" The first one is the question we ask in this chapter for both analyses. When comparing China and Shanxi Province, we consider the regional economy to be the starting point (1) and the national economy to be the end point (2), thus forming an SSDA model.

We can decompose the final-demand shift and production-technology changes. As indicated earlier, we can divide the final-demand shift into the changes in the level, the distribution, and the pattern of demand.

$$Y = \mathbf{MDL} \tag{4.6}$$

Where **M** is the matrix of the spending mix of individual final-demand sectors; **D** is the diagonal matrix with the sector distribution of total demand on the diagonal; and **L** is the diagonal matrix with the overall total demand level on the diagonal (Lin 1996).

$$\Delta Y = Y_2 - Y_1 = \mathbf{M}_2\mathbf{D}_2\mathbf{L}_2 - \mathbf{M}_1\mathbf{D}_1\mathbf{L}_1 \tag{4.7}$$

$$\begin{aligned}
\Delta E &= F_1 (Y_2 - Y_1) + e (Y_2 - Y_1)\, n \\
&= F_1 (\mathbf{M}_2\mathbf{D}_2\mathbf{L}_2 - \mathbf{M}_1\mathbf{D}_1\mathbf{L}_1)\, u + e (\mathbf{M}_2\mathbf{D}_2\mathbf{L}_2 - \mathbf{M}_1\mathbf{D}_1\mathbf{L}_1)\, n \\
&= F_1\mathbf{M}_1\mathbf{D}_1 (\mathbf{L}_2 - \mathbf{L}_1)\, u + e\mathbf{M}_1\mathbf{D}_1 (\mathbf{L}_2 - \mathbf{L}_1)\, n \qquad \textit{(Level effect)} \\
&+ F_1\mathbf{M}_1 (\mathbf{D}_2 - \mathbf{D}_1)\, \mathbf{L}_2\, u + e\mathbf{M}_1 (\mathbf{D}_2 - \mathbf{D}_1)\, \mathbf{L}_2\, n \quad \textit{(Distribution effect)} \\
&+ F_1 (\mathbf{M}_2 - \mathbf{M}_1)\, \mathbf{D}_2\mathbf{L}_2\, u + e (\mathbf{M}_2 - \mathbf{M}_1)\, \mathbf{D}_2\, \mathbf{L}_2\, n \qquad \textit{(Pattern effect)}
\end{aligned} \tag{4.8}$$

Where **u** is a 9x1 matrix whose elements are all equal to 1.

4.2 SDA Results: Shanxi Province, 1992 and 1999

As noted in the previous section, using the energy input-output model, we identify two parts of energy consumption: intermediate and final. In this section, we present the results of the analysis for energy consumption by intermediate consumers, and we show the differences in the amount of energy consumed per unit of output for Shanxi Province in 1992 and 1999.

For the following calculations, we assume the production technology to be that of Shanxi Province in 1992 and perform a set of computations to determine how 1992 to 1999 final-demand shifts affect energy consumption in Shanxi Province. We calculate the impact of final-demand shifts along the three dimensions described previously (level, distribution, and patterns) of final demand. Although the three dimensions overlap, combined they provide a unique insight into the differences in final demand and energy consumption in Shanxi Province between 1992 and 1999. Briefly, we find that in most sectors, the direct inputs of energy resources decrease during this period, but not for electricity, steam & hot water. Furthermore, the total (direct and indirect) energy-input coefficients show that all the sectors consumed less energy in 1999 than in 1992 to produce one unit of output. However, we also find that petroleum refineries & cokemaking are the major contributors to the coal-intensity increase. We conclude that for most sectors, technology improvement helped reduce the use of the most important energy resource, coal, significantly. We discuss each of these findings in detail next.

4.2.1 Level and Distribution of Final Demand

By using SDA, we conduct an in-depth study of energy intensities in Shanxi Province for 1992 and 1999, holding production technology constant. We conduct the analysis for six energy-consumption categories: (1) peasant, (2) non peasant, (3) institutional, (4) gross fixed capital formation (GFCF), (5) change in stocks, and (6) net exports (Tables 4.1 and 4.2).

The total final use of coal, coke products, refined petroleum, and electricity increased by 608.9, 13.4, 760.3, and 1,559.0 million Renminbi (RMB), respectively, from 1992-1999. The only exception in this surge of energy uses is that of crude petroleum & natural gas, which decreased by 654.2 million RMB, although most of this latter is a decrease in stocks of 587.9 million RMB. We do not have direct evidence concerning why the consumption of crude petroleum & natural gas by nonpeasant consumers decreased by 54.2 million RMB between the two years, but the dramatic shift from manual labor who use crude petroleum and natural gas for heating and lighting to machines that generally use other forms of energy, such as electricity and refined petroleum, could be one of the most important reasons.

Although among the final demand sectors, peasants and nonpeasants, as well as total institutional sectors, consumed more energy products in 1999 than in 1992, we note that, as expected, in 1999 peasants spent more on coal (422.6 million RMB) than on electricity (377.6 million RMB) while nonpeasants spent more on electricity (965.7 million RMB) than on coal (232.1 million RMB). This changed since 1992 when both spent more on coal than on electricity, indicating that coal is slowly being replaced by other energy sources in most urban areas in Shanxi Province, which we discuss in detail later in this chapter. Overall, as is apparent from Table 4.3, Shanxi Province consumed significantly more energy in 1999 than 1992, with the one exception of crude petroleum and natural gas.

The distribution coefficients of final demand for the five final-demand sectors vary considerably (Tables 4.1-4.3). In 1992, the final demand for coal came primarily from peasant and nonpeasant consumption, which accounted for 62.9% and 28.2% of the total final demand, respectively. However, by 1999, the demand for coal by final demand sectors was spread more evenly across peasant, nonpeasant, and institutional consumers, as well as changes in stock, accounting for 39.6%, 21.7%, 17.4%, and 21.3% of the total final demand, respectively.

The consumption patterns of electricity and refined petroleum did not change as significantly as that of coal from 1992 to 1999. In 1992, the final demand for electricity was mainly from peasants, nonpeasants, and institutional consumption, which accounted for 33.8%, 58.3%, and 7.9% of the total final demand, respectively; while in 1999, the distribution changed to 22.3%, 57.1%, and 20.6% of the total final demand, respectively. Similarly, in 1992, the final demand for refined petroleum was from peasants, nonpeasant, and institutional consumption, as well as changes in stocks, which accounted for 8.3%, 12.4%, 2.6%, and 76.7% of the total final demand, respectively;

Table 4.1: Energy Requirements of Final Demand Sectors in Shanxi Province, 1992, Absolute Value and Distribution Coefficients

Unit: million RMB

Amount	Coal	Coke & Coal Gas	Crude Petroleum & Natural Gas	Refined Petroleum	Electricity
Peasant Consumption	288.5	0.0	9.6	19.2	45.0
Nonpeasant Consumption	129.6	0.0	54.2	28.7	77.6
Total Institutional Consumption	0.2	0.0	2.5	5.9	10.5
Gross Fixed Capital Formation	0.0	0.0	0.0	0.0	0.0
Change in Stocks	40.6	-2.6	639.5	177.1	0.0
Total Final Use	458.9	-2.6	705.8	231.0	133.1

Unit: direct energy input/thousand RMB of total final demand

Distribution Coefficients	Coal	Coke & Coal Gas	Crude Petroleum & Natural Gas	Refined Petroleum	Electricity
Peasant Consumption	62.9	0.0	1.4	8.3	33.8
Nonpeasant Consumption	28.2	0.0	7.7	12.4	58.3
Total Institutional Consumption	0.04	0.0	0.4	2.6	7.9
Gross Fixed Capital Formation	0.0	0.0	0.0	0.0	0.0
Change in Stocks	8.9	100.0	90.5	76.7	0.0
Total Final Use	100.0	100.0	100.0	100.0	100.0

Source: Calculated from "1992 Shanxi Province input-output table". *1992 Shanxi Statistical Yearbook.* China Statistics Press.

Note: RMB: Renminbi, the Chinese currency, which is equivalent to about 0.125 US Dollar.
Numbers do not add up to the total due to rounding.

Table 4.2: Energy Requirements of Final Demand Sectors in Shanxi Province, 1999, Absolute Value and Distribution Coefficients

Unit: million RMB

Amount	Coal	Coke & Coal Gas	Crude Petroleum & Natural Gas	Refined Petroleum	Electricity
Peasant Consumption	422.6	0.0	0.0	104.7	377.6
Nonpeasant Consumption	232.1	0.0	0.0	257.3	965.7
Total Institutional Consumption	185.9	0.0	0.0	94.2	348.8
Gross Fixed Capital Formation	0.0	0.0	0.0	0.0	0.0
Change in Stocks	227.2	10.8	51.6	535.0	0.0
Total Final Use	1,067.8	10.8	51.6	991.2	1,692.1

Unit: direct energy input/thousand RMB of total final demand

Distribution Coefficients	Coal	Coke & Coal Gas	Crude Petroleum & Natural Gas	Refined Petroleum	Electricity
Peasant Consumption	39.6	0.0	0.0	10.5	22.3
Nonpeasant Consumption	21.7	0.0	0.0	26.0	57.1
Total Institutional Consumption	17.4	0.0	0.0	9.5	20.6
Gross Fixed Capital Formation	0.0	0.0	0.0	0.0	0.0
Change in Stocks	21.3	100.0	100.0	54.0	0.0
Total Final Use	100.0	100.0	100.0	100.0	100.0

Source: Calculated from "1992 Shanxi Province input-output table". *1992 Shanxi Statistical Yearbook.* China Statistics Press.

Note: RMB: Renminbi, the Chinese currency, which is equivalent to about 0.125 US Dollar.
Numbers do not add to the total due to rounding.

Table 4.3: Absolute Difference (1999-1992) in Energy Requirements of Final Demand Sectors in Shanxi Province, Absolute Value and Distribution Coefficients

Unit: million RMB

Amount	Coal	Coke & Coal Gas	Crude Petroleum & Natural Gas	Refined Petroleum	Electricity
Peasant Consumption	134.1	0.0	-9.6	85.5	332.6
Nonpeasant Consumption	102.5	0.0	-54.2	228.6	888.1
Total Institutional Consumption	185.7	0.0	-2.5	88.3	338.3
Gross Fixed Capital Formation	0.0	0.0	0.0	0.0	0.0
Change in Stocks	186.6	13.4	-587.9	357.9	0.0
Total Final Use	608.9	13.4	-654.2	760.3	1559.0

Unit: direct energy input / thousand RMB of total final demand

Distribution Coefficients	Coal	Coke & Coal Gas	Crude Petroleum & Natural Gas	Refined Petroleum	Electricity
Peasant Consumption	-232.9	0.0	-13.6	22.5	-114.9
Nonpeasant Consumption	-65.1	0.0	-76.8	135.3	-12.3
Total Institutional Consumption	173.7	0.0	-3.5	69.5	127.2
Gross Fixed Capital Formation	0.0	0.0	0.0	0.0	0.0
Change in Stocks	124.3	0.0	93.9	-227.2	0.0
Total Final Use*	0.0	0.0	0.0	0.1	0.0

Source: Calculated from Tables 4.1-4.2.
RMB: Renminbi, the Chinese currency, which is equivalent to about 0.125 US Dollar.
*Note that these should all total to zero, but do not due to rounding.

while in 1999, the percentages changed to 10.5%, 26.0%, 9.5%, and 54.0% of the total final demand, respectively. In both years, the change in stocks accounted for more than half of the total final consumption.

The distribution of public and private consumption of energy products by peasants, nonpeasants, and institutional consumers in Shanxi Province is changing over time. We analyze this to see how this change in distribution influences the amount of energy used. As shown in Tables 4.1 and 4.2, peasants were the major coal consumers in both 1992 and 1999 among the public and private consumers, while nonpeasant consumers consumed most of the electricity, petroleum, coke, and other coal products. In other words, in rural areas, people tend to use more raw coal as an energy source; in contrast, people in urban areas tend to use processed energy resources.

Although coal traditionally has been the dominant source of energy in Shanxi Province, the province today tends to use different kinds of energy sources. However, because petroleum is scarce and is higher in price than coal in Shanxi Province, this could raise the cost of energy use in Shanxi Province. For example, in 1994, residents of Taiyuan City, Shanxi Province, used a widely distributed type of gasoline, called gasoline 70, which cost 2,780 RMB/tonne, while bituminous coal cost only 195 RMB/tonne (Sinton 1996: VI-8). Even if we take the standard coal equivalent into account, the cost of petroleum is still seven times greater than bituminous coal. Another significant phenomenon is that the percentage of coke, gas, and coal products consumption of total energy consumption decreased, indicating that peasants and nonpeasants were switching from coal and gas to other forms of energy for residential use.

4.2.2 Change in the Direct Energy-Input Coefficients

For this part of the analysis, we separate the 33 sectors in the input-output table into two groups: energy-production sectors and nonenergy-production sectors. Table 4.4 shows the difference between 1992 coefficients and 1999 coefficients (1999 minus 1992), from which we draw three conclusions.

First, when comparing the total of the energy direct-input coefficients for all 33 sectors, we note that the coefficient is lower in 1999 for three of the five energy resources: crude petroleum, coke & coal gas, and refined petroleum. The most significant reduction in energy direct inputs per thousand units of output is in refined petroleum (-323.9), followed by coke & coal gas (-114.7), followed by a negligible reduction for crude petroleum (-0.3).

Second, although the coal direct inputs increased by 69.9 per thousand units of output, that amount is significantly inflated by the 609.0 increase of coal to produce refined petroleum (100.9) and coke products (508.1). As such, without taking into account this anomaly in the coal direct inputs, the total coal direct inputs for the remaining 31 sectors is almost 540 units of coal per thousand units of output lower in 1999 than in 1992. Because of the importance of coal as an energy resource in Shanxi Province, this impressive reduction in coal direct inputs for the remaining 31 sectors is an important accomplishment in increased energy efficiency in Shanxi Province.

Just as important, however, is the large increase in coal direct inputs into the coke and refined petroleum industries. Although it is difficult to pinpoint exactly the cause of this increase, there are two possible explanations. The first is the lowering of the quality of coking coal as coke production soared during the 1990s in Shanxi Province, which resulted in coke plants requiring more coal input to produce the same amount of coke output, i.e., more poor-quality coal is needed to produce the same amount of coke as could be produced if the coal were of higher quality. The second is that this increase in demand for Shanxi Province coke resulted in the Township and Village Enterprises (TVEs) establishing and operating a number of smaller cokemaking plants, which were not as energy efficient as the cokemaking plants that were State-Owned Enterprises (SOEs), i.e., TVE plants generally use more coal to produce a tonne of coke than SOE plants (Chapter 2). In fact, as of 2004, many of these small TVE cokemaking plants have been closed for efficiency and environmental reasons.

The third conclusion from Table 4.4 is that unlike other energy inputs, electricity direct-inputs increased for 24 of the 33 sectors, for an aggregate increase of 208.3 units of electricity inputs per thousand units of output. Not only is this increase in stark contrast to the decreases in the remaining four energy inputs, we are surprised to find such a widespread decrease in electricity efficiency across so many sectors. Among the 24

Table 4.4: Difference between 1992 and 1999 Energy Direct-Input Coefficients in Shanxi Province

Unit: energy direct input/thousand units of output

Sector	Coal	Coke & Coal Gas	Crude Petroleum & Natural Gas	Refined Petroleum	Electricity
Energy Sectors					
Coal	-6.6	-1.1	0.0	-10.6	-0.3
Coke & Coal Gas	508.1	-29.0	0.0	-12.5	-11.0
Crude Petroleum & Natural Gas	-23.5	0.0	0.0	2.2	14.4
Refined Petroleum	100.9	-0.6	0.0	-258.4	-34.8
Electricity	-54.3	-0.1	0.0	-6.9	20.4
NonEnergy Sectors					
Agriculture	-3.2	-2.1	0.0	-0.5	-6.0
Metal Ores Mining	-11.0	-0.1	0.0	-14.8	71.3
Other Mining	-19.9	-3.8	0.0	-55.9	-48.8
Food	-11.2	-0.2	0.0	-1.3	1.9
Textiles	-8.8	-0.2	0.0	-3.4	5.8
Apparel Products	-1.8	0.0	0.0	2.6	24.2
Sawmills	-50.8	-0.7	0.0	-6.8	18.2
Paper	-24.0	-0.2	-0.3	-1.2	6.1
Chemicals	-12.0	-19.2	0.0	5.5	12.5
Building Materials	-43.0	-19.9	0.0	47.0	-13.5
Primary Metal	-40.0	26.6	0.0	46.9	12.9
Metal Products	-11.1	-9.0	0.0	-10.5	4.9
Machinery	-22.8	-2.8	0.0	1.9	-5.3
Transport Equipment	-11.9	-2.1	0.0	-2.3	3.1
Electric Machinery	-0.5	-3.6	0.0	-0.3	2.8
Electronic Equipment	1.0	0.0	0.0	0.8	16.2
Instruments & Meters	-10.0	-0.6	0.0	-2.3	31.0
Maintenance of Equipment	-22.2	-5.9	0.0	0.9	8.4
Other Industries	-32.6	-21.6	0.0	-8.4	30.4
Construction	-5.2	0.0	0.0	-0.1	-2.2
Freight Transport	-10.3	0.8	0.0	-30.4	10.8
Commerce	-15.6	-2.0	-0.2	-9.0	10.5
Restaurants	-12.2	0.0	0.0	4.3	2.6
Passenger Transport	-10.9	-0.1	0.0	52.8	7.0
Public Utilities	-18.8	-15.2	0.0	-17.0	8.0
Research Institutions	-12.5	-1.9	0.0	11.0	-2.3
Finance & Insurance	-11.9	0.0	0.0	-17.9	15.9
Public Admin.	-21.5	0.0	0.0	-7.4	-7.0
Total	69.9	-114.7	-0.3	-323.9	208.3

Source: 1992 and 1999 Shanxi Province input-output tables. *1992 Shanxi Statistical Yearbook.* China Statistics Press. *1999 Shanxi Statistical Yearbook.* China Statistics Press.

Note: Numbers do not add to the total due to rounding.

sectors where the electricity direct inputs increased, metal-ore mining showed the biggest increase (71.3 units of electricity per 1000 units of output), and of the 9 sectors where the electric direct inputs decreased, the other mining sector exhibited the greatest decline (-48.8 units of electricity per 1000 units of output).

4.2.3 Differences in Energy Direct-and-Indirect (Total) Input Coefficients

Table 4.5 shows the differences in direct-and-indirect (total) inputs of energy in Shanxi Province between 1992 and 1999. Because indirect energy use represents a much bigger

Table 4.5: Difference between 1992 and 1999 Energy Direct-and-Indirect Input Coefficients, Shanxi Province

Unit: direct & indirect input/thousand units of output

Sector	Coal	Coke & Coal Gas	Crude Petroleum & Natural Gas	Refined Petroleum	Electricity
Energy Sectors					
Coal	-27.9	-314.9	-115.4	-396.5	-25.2
Coke & Coal Gas	1102.9	-34.4	459.4	-491.3	95.4
Crude Petroleum & Natural Gas	-206.1	-297.9	0.0	-218.5	180.3
Refined Petroleum	101.7	-671.2	-519.1	-272.2	-493.8
Electricity	-116.7	21.0	-198.9	-218.2	29.7
Non-Energy Sectors					
Agriculture	-226.8	-539.3	-855.3	-22.3	27.9
Metal Ore Mining	148.8	12.4	-379.8	-323.9	791.1
Other Mining	-424.8	-104.6	-534.8	-629.9	-324.1
Food	-273.5	-322.8	-798.7	-160.6	195.1
Textiles	-249.1	-459.3	-179.9	-353.5	187.5
Apparel Products	-110.1	-386.7	-364.7	-156.0	590.6
Sawmills	-514.8	-593.7	-697.5	-461.0	112.2
Paper	-351.1	-423.0	-937.0	-277.0	127.3
Chemicals	-164.4	-645.9	-519.4	-65.2	273.4
Building Materials	-257.5	-518.2	-28.2	399.8	6.5
Primary Metal	135.1	341.4	177.6	634.6	393.5
Metal Products	85.4	162.3	-113.5	231.6	474.3
Machinery	-58.3	361.0	-5.1	309.1	268.3
Transport Equipment	-212.9	-22.2	194.4	-56.7	139.8
Electric Machinery	-62.6	-139.9	-358.0	150.1	258.6
Electronic Equipment	-152.4	-90.4	-727.9	-181.2	349.1
Instruments & Meters	206.6	350.7	-341.0	249.2	1085.4
Maintenance of Equipment	-84.0	273.0	-267.1	110.9	321.7
Other Industries	-404.1	-720.7	-607.5	-327.1	209.4
Construction	-191.2	-314.4	-327.5	-21.1	84.4
Freight Transport	42.4	286.6	-393.4	-308.8	356.9
Commerce	-288.3	-300.1	-928.5	-280.1	336.3
Restaurants	-239.7	148.2	-748.4	-64.0	277.4
Passenger Transport	231.4	280.9	-243.6	256.5	536.4
Public Utilities	-138.8	-421.8	-583.7	-179.5	613.3
Research Institutions	-330.3	-412.8	-317.4	365.1	70.8
Finance & Insurance	-255.4	-324.9	41.8	-482.7	426.0
Public Admin.	-560.6	-552.4	-869.9	-490.4	-213.1
Totals	-3847.1	-6374.0	-12088.0	-3730.8	7762.4

Source: 1992 and 1999 Shanxi Province input-output tables. *1992 Shanxi Statistical Yearbook*. China Statistics Press. *1999 Shanxi Statistical Yearbook*. China Statistics Press.
Note: Numbers do not add to total due to rounding.

share of the total energy used in the economy than the energy used by final users, the results of this analysis are a good reflection of the changes in energy inputs in Shanxi Province between 1992 and 1999. We have two findings.

First, with the exception of electricity, these energy direct-and-indirect (total) inputs for most of the sectors decreased during the 1990s in contrast to the direct input trend. For the coal, coke & coal gas, crude petroleum & natural gas, refined petroleum, and electricity sectors, the total input coefficients decreased for 25, 23, 28, 24, and 4 sectors out of 33 sectors, respectively. Similar to the energy direct inputs, for most sectors (29 out of 33), the electricity total inputs increased between 1992 and 1999.

Second, concerning the coal direct-and-indirect (total) inputs, we noted earlier that there were significant increases in coal direct inputs into the cokemaking and refined petroleum sectors between 1992 and 1999, which resulted in an aggregate increase in direct coal inputs between 1992 and 1999. In fact, coal total inputs also increase significantly for the two sectors between 1992 and 1999, but unlike coal direct inputs, the aggregate reduction in coal total inputs for the remaining 31 sectors results in an overall decrease in coal total inputs. This is impressive because the coal total inputs to the cokemaking and refined petroleum sectors increased by over 1,200 units of coal per 1,000 units of output, implying that the total reduction for all the sectors was almost 4,500 units of coal per 1,000 units of output. Finally, for 9 of the 33 sectors, the energy total input increased between 1992 and 1999, most of which came from additional electricity inputs. This compares to 7 of the 33 sectors where the energy direct inputs increased between 1992 and 1999, again, primarily due to increased electricity direct input. Thus, sectors that rely heavily on electricity as an energy input saw their overall energy inputs increase between 1992 and 1999.

4.2.4 Source of the Energy-Use Change

With SDA, we studied the source of the energy-use change. As shown in Table 4.6, Eg is the vector of intermediate energy consumption, Ed is the vector of final energy consumption, and the total energy consumption in the economy, E, is the sum of intermediate and final energy consumption.

In both 1992 and 1999, intermediate consumption was usually lower than final consumption for coke, gas, & coal products, but higher for crude petroleum & natural gas, refined petroleum, and electricity. However, the intermediate consumption of crude petroleum, electricity, and refined petroleum was significantly higher than the direct energy consumption. The intermediate consumption (Table 4.6) for all categories of energy products increased in 1999 compared with 1992, ranging from 33.7% (coal mining) to 209.9% (electricity, steam, and hot water). The direct-energy consumption for coal mining and cokemaking, gas, & coal products decreased dramatically in 1999 compared with 1992, while for the other three sectors, the direct-energy consumption increased from 1992 to 1999.

We use the SDA framework to separate the energy-use changes into two parts: the final-demand shift and the production-technology change. The final-demand shift indicates the energy impact of final-demand changes while production technology is held constant. The production-technology change quantifies the energy effect of changes in production technology with a given final demand. Table 4.7 shows the significance of each of these parts in the overall changes in the consumption of all energy products.

Table 4.6: Intermediate Energy Consumption (Eg), Final Energy Consumption (Ed), and Total Energy Consumption (E), Shanxi Province, 1992 and 1999
(Million RMB, 1992 Prices, and Percent Difference)

Sector	Intermediate Energy Consumption (E_g) 1992 (Million RMB)	1999 (Million RMB)	Percent Change 1999 - 1992 (Percent)
Coal Mining	6,144.6	15,061.7	145.1
Cokemaking, Gas, & Coal Products	1,244.1	2,825.7	127.1
Crude Petroleum & Natural Gas	4.3	5.8	33.7
Petroleum Refineries	2,749.4	8,648.7	214.6
Electricity, Steam, & Hot Water	3,885.9	12,040.6	209.9

Sector	Direct Energy Consumption (E_d) 1992 (Million RMB)	1999 (Million RMB)	Percent Change 1999 - 1991 (Percent)
Coal Mining	12,061.7	1,067.8	-91.2
Cokemaking, Gas, & Coal Products	2,467.4	51.6	-97.9
Crude Petroleum & Natural Gas	-2.6	10.8	-510.4
Petroleum Refineries	233.4	991.1	324.6
Electricity, Steam, & Hot Water	611.4	1,692.1	176.8

Sector	Total Energy Consumption (E) 1992 (Million RMB)	1999 (Million RMB)	Percent Change 1999 - 1992 (Percent)
Coal Mining	18,206.3	16,129.4	-11.4
Cokemaking, Gas, & Coal Products	3,711.5	2,877.3	-22.5
Crude Petroleum & Natural Gas	1.7	16.6	885.5
Petroleum Refineries	2,982.9	9,639.8	223.2
Electricity, Steam, & Hot Water	4,497.3	13,732.7	205.4

Source: SDA calculation from 1992 and 1999 Shanxi input-output table. *1992 Shanxi Statistical Yearbook.* China Statistics Press. *1999 Shanxi Statistical Yearbook.* China Statistics Press.
Note: RMB: Renminbi, the Chinese currency, which is equivalent to about 0.125 US Dollar.

Table 4.7: Energy-Use Change and the Sources of the Change, Structural Decomposition Analysis

Sector	Energy-Use Change (million RMB)	Final Demand Shift (million RMB)	Final Demand Shift (% Change)	Production Technology Change (million RMB)	Production Technology Change (% Change)
Coal Mining	-5,504.8	-5,003.6	90.9%	-501.2	9.1%
Cokemaking, Gas, & Coal Products	-1,266.7	-828.9	65.4%	-437.7	34.6%
Crude Petroleum & Natural Gas	14.1	17.1	121.0%	-3.0	-21.0%
Petroleum Refineries	4,369.5	5,324.3	121.9%	-954.8	-21.9%
Electricity, Steam, & Hot Water	6,417.1	5,394.3	84.1%	1,022.9	15.9%

Source: SDA calculation from "1992 Shanxi Input-Output Table" and "1999 Shanxi Input-Output Table." *1992 Shanxi Statistical Yearbook.* China Statistics Press. *1999 Shanxi Statistical Yearbook.* China Statistics Press.
Note: RMB: Renminbi, the Chinese currency, which is equivalent to about 0.125 US Dollar.

As is apparent from Table 4.7, the final-demand shift of coal mining; cokemaking, gas, and coal products; crude petroleum & natural gas; petroleum refineries; and electricity, steam, & hot water accounted for 90.9%, 65.4%, 121.0%, 121.9%, and 84.1% of the energy-use changes, respectively, while production-technology changes accounted for 9.1%, 34.6%, -21.0%, -21.9%, and 15.9%, respectively. In fact, production-technology changes helped Shanxi Province to consume significantly lower amounts of coal, crude petroleum, coke, and refined petroleum, which points to great reductions in energy intensity, except for electricity where the production-technology change of electricity is positive, indicating that, in this sector, energy intensity actually increased.

Table 4.8: Structural Decomposition Analysis of Final Demand for Energy Sectors in Shanxi Province, 1992 and 1999

Unit: million RMB

Total Amount	Final-Demand Shift	Level Effect	Distribution Effect	Pattern Effect
Coal Mining	-5003.1	17502.8	-18337.2	-4168.6
Cokemaking, Gas, & Coal Products	-828.9	3749.2	-2642.5	-1935.6
Crude Petroleum & Natural Gas	17.1	3.	0.7	13.3
Petroleum Refineries	5325.1	5204.2	2688.8	-2568.0
Electricity, Steam, & Hot Water	5394.9	5839.4	1519.6	-1964.0

Unit: percent

Percentage of Total	Final Demand Shift	Level Effect	Distribution Effect	Pattern Effect
Coal Mining	100.0	-350.0	366.5	83.3
Cokemaking, Gas, & Coal Products	100.0	-452.3	318.8	233.5
Crude Petroleum & Natural Gas	100.0	18.1	3.9	78.0
Petroleum Refineries	100.0	97.7	50.5	-48.2
Electricity, Steam & Hot Water	100.0	108.2	28.2	-36.4

Source: SDA calculation from "1992 Shanxi Input-Output Table" and "1999 Shanxi Input-Output Table." *1992 Shanxi Statistical Yearbook.* China Statistics Press. *1999 Shanxi Statistical Yearbook.* China Statistics Press.

Note: RMB: Renminbi, the Chinese currency, which is equivalent to about 0.125 US Dollar.

Finally, we divided the final-demand shift into the three components (level, distribution, and pattern of demand). Table 4.8 shows the energy effects of final-demand distribution and pattern changes between 1992 and 1999. We use 1992 as the base year, and 1999 as the year under consideration. The table shows how much more Shanxi Province consumed as final products in 1999 than in 1992, and what are the individual contributions of distribution and pattern effects. We draw three conclusions from this table. First, the level effect, i.e., the increase in provincial GDP, creates the largest final-demand increase in all the five energy sectors. For all of these final demand sectors, level effects are positive: 17,502.8 million RMB (coal), 3.1 million RMB (Crude Petroleum & Natural Gas), 5,839.4 million RMB (Electricity, Steam, & Hot Water), 5,204.2 million RMB (Petroleum Refineries), and 3,749.19 million RMB (Cokemaking, Gas, & Coal Products), respectively. Second, the distribution effect and pattern effect played a very important role in two sectors: coal mining, and cokemaking, gas, & coal products; in fact, the decrease of these two effects even offset the large increase of the level effect, making the final-demand shift of the two sectors decrease from 1992 to 1999. The third conclusion is the different role that the distribution effect plays in different sectors. It helped to reduce the final demand in coal mining and cokemaking, but to increase the final demand in the other three energy sectors. In summary, Table 4.8 shows that the major contributor is the level effect, which together with the pattern effect always helps to reduce the final energy consumption, but the distribution effect plays different roles in different sectors.

In all, using SDA, we can separate the changes of energy-use in the economy into final demand changes and/or production technology. As already indicated, we also decomposed the final-demand-shift components into changes in the level, distribution, and pattern of final demand.

4.3 SSDA Results: Shanxi Province and China in 1995

We conducted a similar analysis as above, SSDA, using the same methods for China and Shanxi Province for 1995. For these calculations, we assume the production technology to be that of China as a whole and perform a set of computations using China's pattern and distribution of final demand for Shanxi Province to determine their effects on energy consumption in Shanxi Province's economy. We calculate the impact of final-demand shifts for two (distribution and pattern) of the three dimensions of final demand used previously. Normally, when conducting SDA on a temporal level, as was the case above, analysts would also take level patterns into consideration. This is because final users may buy more or less of everything and keep the mix of goods and services they purchase constant. However, this is not a relevant consideration in this case, because we are comparing China as a whole to one of her provinces, where there is a significant difference in the level of final demand. As such, in this case, we omit the level effect from the final-demand shifts.

As we show in this section, the primary factors behind the differences in energy-intensity levels between Shanxi Province and China are due to differences in direct and indirect energy inputs, especially coal, among the 33 sectors under observation. Thus, our analysis shows that the reason for the additional energy inputs is due to differences in production technologies between China and Shanxi Province. We also show that shifts in final demand play a small role in explaining the overall difference, the majority of which were due to the fact that the five final-demand sectors consumed more energy-intensive products in Shanxi Province than in China. Finally, a negligible percentage can be attributed to the fact that the industrial composition in these five sectors was different between Shanxi Province and China. We discuss each of these factors in detail next.

4.3.1 Distribution and Pattern of Final Demand

Table 4.9 shows the difference in energy requirements of final-demand sectors for Shanxi Province and China.

Table 4.9 shows that Shanxi Province relies on coal more heavily than China for peasant and nonpeasant consumption and less on electricity. Another significant difference is the changes in stocks between Shanxi Province and China. The State Statistical Bureau defines changes in stocks as being equal to the value of stocks acquired, minus the value of stocks disposed of in 1995 (1995 Input-Output Table of China). There is a significant difference between the two regions in this final-demand category, particularly in coke & coal gas. This makes sense considering that Shanxi Province is the largest coke producer in China, the majority of which is exported to other regions in China, and overseas.

In addition to the distribution of final demand, the pattern of final demand, that is, the mix of goods and services within an individual final-demand sector, is different between Shanxi Province and China. Energy consumption, for instance, can increase if there is a shift in the spending pattern from less energy-intensive commodities, such as commerce and services, to more energy-intensive ones, such as chemical fertilizers and cement. To display the results of our analysis in Table 4.10, we use a monetary ratio of RMB of energy-use per thousand RMB of GDP. In the table, a positive number indicates that energy intensity in China is larger than in Shanxi Province, while a negative number indicates the opposite. Table 4.10 shows that the mix of goods and services is different between Shanxi Province and China in our final-demand sectors: peasant, nonpeasant, and institutional consumption along with gross fixed capital formation and changes in stocks.

Table 4.11 shows the summary results of the importance of final-demand distribution and pattern changes between Shanxi Province and China.

There are three facts that stand out from Table 4.11. First, and most important, is the fact that there is a difference in every final-demand category under consideration among all fuel types. The difference in total effects is greatest for coal, 5.0%, and negative for crude petroleum & natural gas at minus 0.1%. This is perfectly plausible given Shanxi Province's heavy reliance on coal. The second fact is the importance of the pattern effect in the overall final-demand-shift (total effect) between China and Shanxi Province. Over 90% of the final-demand shift can be explained by the fact that final consumers in Shanxi Province consume more of the products that are more energy intensive than do the final users in China as a whole. The third fact is the insignificant contribution of the distribution effect to the total effect. Altogether the distribution effect contributed only 0.5% to the energy-consumption differences between Shanxi Province and China.

In summary, Table 4.11 shows that in the absence of production-technology differences, which we discuss next, final-demand differences between Shanxi Province and China were responsible for only 12.8 percent of the differences in energy consumption in monetary values (RMB). In addition, almost all of this was due to the differences in the mix of goods and services within the individual demand sectors. Furthermore, because there are large differences in product energy intensities, these changes in the mix of goods and services purchased by final consumers did have an impact on the overall difference in energy consumption between Shanxi Province and China.

Table 4.9: Differences in Energy Requirements of Final Demand Sectors, China and Shanxi Province, 1995

Unit: RMB of energy use/thousand RMB of final demand

Sector	Coal	Coke & Coal Gas	Crude Petroleum & Natural Gas	Refined Petroleum	Electricity
Peasant Consumption	14.7	-1.7	0.0	0.7	-4.1
Nonpeasant Consumption	9.0	-1.4	-0.1	1.2	-3.2
Institutional Consumption	0.0	0.3	0.0	0.8	1.4
Gross Fixed Capital Formation	0.0	0.0	0.0	0.0	0.0
Changes in Stock	6.3	99.1	-21.2	36.8	0.0

Source: SDA calculated from "1995 Shanxi Input-Output Table". *1995 Shanxi Statistical Yearbook.* China Statistics Press.
Note: RMB: Renminbi, the Chinese currency, which is equivalent to about 0.125 US Dollar.

Table 4.10: Differences In the Spending Mix of Selected Final-Demand Sectors Between Shanxi Province and China, 1995

Unit: percent

	Energy	Agriculture	Industry	Transport	Construction	Services
Peasant Consumption	-8.2	-2.7	-4.2	0.0	0.1	15.0
Nonpeasant Consumption	-11.3	-4.2	7.3	0.0	0.2	7.9
Institutional Consumption	1.0	-1.9	-0.3	0.0	-0.0	1.2
Gross Fixed Capital Formation	9.0	2.6	10.4	-26.5	3.0	1.5
Changes in Stocks	-0.5	7.3	-0.8	0.0	3.5	-9.6
Total	-10.0	1.2	12.4	-26.5	6.8	16.1

Source: SSDA calculations from "1995 China and Shanxi Province Input-Output Tables." *1995 Shanxi Statistical Yearbook.* China Statistics Press. *State Statistical Bureau of China (SSB) 1995. China Input-Output Table, 1995 (English Edition).* Beijing: China Statistical Press.
Note: Numbers do not add to total due to rounding.

Table 4.11: Primary Energy-Use Differences between Shanxi Province and China: Final-Demand Shift, Distribution, and Pattern Changes, 1995

Unit: percent difference with China

Effect	Coal	Coke & Coal Gas	Crude Petroleum & Natural Gas	Refined Petroleum	Electricity	Total
Distribution Effect	-0.2	0.6	-0.1	0.2	0.0	0.5
Pattern Effect	5.1	1.7	-0.0	3.6	1.8	12.2
Total	5.0	2.3	-0.2	3.9	1.8	12.8

Source: SSDA calculations from 1995 China and Shanxi Province Input-Output Tables. *1995 Shanxi Statistical Yearbook.* China Statistics Press. *State Statistical Bureau of China (SSB). 1995. China Input-Output Table, 1995 (English Edition).* Beijing: China Statistical Press.
Note: Numbers do not add to total due to rounding.

4.3.2 Importance of Energy Direct Inputs

Energy direct-input requirements differ significantly between Shanxi Province and China (Table 4.12). First, and most important, is the difference in coal requirements in terms of energy direct inputs between the two regions. In fact, coal mining and coke production are the only two sectors where the coal direct-input requirement is smaller in Shanxi Province than in China, and the differences in these two sectors are very small.

Table 4.12: Differences in Direct Energy Requirements, Shanxi Province Minus China, 1995

Unit: direct energy inputs/thousand RMB of output

Sector	Coal	Coke & Coal Gas	Crude Petroleum & Natural Gas	Refined Petroleum	Electricity
Energy sectors					
Coal	-13.7	1.0	0.1	-3.4	-16.0
Coke & Coal Gas	28.4	0.6	-521.2	279.5	68.6
Crude Petroleum & Natural Gas	0.0	0.0	0.0	0.0	0.0
Refined Petroleum	88.4	0.0	-16.1	14.9	40.7
Electricity	237.0	-0.1	-37.7	-30.9	-11.4
NonEnergy Sectors					
Agriculture	6.4	2.0	0.0	-1.7	11.4
Metal Ores Mining	1.7	0.5	-0.1	33.3	1.4
Other Mining	11.9	3.7	-0.2	56.2	17.1
Food	14.1	-0.3	0.0	-0.2	4.0
Textiles	13.8	0.1	0.0	2.8	15.8
Apparel Products	5.7	-0.2	0.0	0.6	4.7
Sawmills	60.5	0.3	0.0	1.2	5.7
Paper	35.6	-0.2	0.1	2.9	14.5
Chemicals	27.9	18.8	-15.3	-0.6	16.5
Building Materials	79.1	20.4	-3.3	-5.2	13.4
Primary Metal	30.9	39.0	-5.5	9.2	0.7
Metal Products	14.9	7.5	-0.3	10.8	7.7
Machinery	28.5	2.2	-0.5	3.9	9.9
Transport Equipment	0.0	0.0	.0	0.0	0.0
Electric Machinery	17.9	1.8	-0.3	3.1	20.5
Electronic Equipment	6.4	3.4	-0.2	5.0	3.3
Instruments & Meters	0.0	0.0	0.0	0.0	0.0
Maintenance of Equipment	4.5	-0.4	-1.0	0.2	-1.0
Other Industries	0.0	0.0	0.0	0.0	0.0
Construction	15.7	-0.4	-1.3	8.6	14.9
Freight Transport	0.0	0.0	0.0	0.0	0.0
Commerce	25.2	-1.4	-1.4	14.4	5.3
Restaurants	33.0	18.7	-2.6	9.3	28.0
Passenger Transport	5.9	0.2	0.0	13.5	7.3
Public Utilities	5.1	0.0	-2.5	24.1	1.7
Research Institutions	16.4	1.6	-0.6	9.4	9.7
Finance & Insurance	21.5	-2.0	-2.9	-2.9	3.6
Public Administration	0.0	0.0	0.0	0.0	0.0

Source: SSDA calculations from "1995 China and Shanxi Province Input-Output Tables." *1995 Shanxi Statistical Yearbook*. China Statistics Press. *State Statistical Bureau of China (SSB). 1995. China Input-Output Table, 1995 (English Edition)*. Beijing: China Statistical Press.

Concerning the coal direct input, the additional 237.0 units of direct coal utilization to produce 1,000 units of electricity had the most significant impact on coal consumption for the five energy sectors. In the nonenergy sectors, the industrial sectors, such as sawmills (60.5) and primary metals (30.9), contributed significantly to the overall differences, in terms of coal direct inputs. Among the other energy inputs, the two that stand out are the coke inputs into the primary-metal sector (39.0), and the direct input of refined petroleum into the other-mining sector (56.2). At the same time, for a number of sectors, no differences exist between the direct energy inputs between Shanxi Province and China. This is important, because it indicates that Shanxi Province policy makers need to concern themselves with only a few strategic economic sectors in order to reduce energy-intensity levels and bring the overall levels as low as those in China as a whole.

4.3.3 Differences in Direct-and-Indirect (Total) Energy Inputs Coefficients

Table 4.13 shows the differences in direct-and-indirect energy inputs between Shanxi Province and China. We find the following four important facts concerning these differences.

First, the differences in energy direct inputs and direct-and-indirect inputs between Shanxi Province and China are significant. The combined difference for the five energy sectors is 71% (890–256/890) greater when we take into account the indirect inputs, the majority of which comes from the coal sector (55%). The difference in the nonenergy sectors is similarly large. In the coal sector, the combined difference (2,545-567/2,545) is over 74%, and, in the refined petroleum sector, it is over 85% (1,321-188/1,321). Altogether, the combined nonenergy sectoral difference (4,551-1081/4,551) is over 76%, and the difference (5,441-1,122/5,441) for energy and nonenergy combined is over 79% greater. The size of these differences suggests that energy indirect inputs per unit of output are very important.

Second, once again, the differentials in coal direct-and-indirect inputs per unit of output between Shanxi Province and China are large. In the five energy sectors alone, Shanxi Province consumes 55% (745-332/745) more coal in terms of direct-and-indirect inputs than China, and combined with the 28 nonenergy sectors, that number jumps to over 73% (3,290-899/3,290). This same tendency holds true for the electricity, refined petroleum, and coke & coal products, where the energy direct-and-indirect energy input in Shanxi Province is 75% greater than that of China (Table 4.13).

Third, the differences among individual energy and nonenergy sectors are also important. The industrial sectors show the largest differences, highlighted by the building materials, sawmills, and furniture sectors. At the same time, there is a significant difference between Shanxi Province and China in the amount of refined petroleum used by the transportation sectors, especially freight transport. We also note that overall, there is one magnitude of difference between the difference in coal direct-and-indirect inputs and any of the other fuel inputs.

Table 4.13: Differences in Direct and Indirect Energy Requirements, Shanxi Province Minus China, 1995

Unit: energy direct and indirect inputs/thousand units of output

Sector	Coal	Coke & Coal Gas	Crude Petroleum & Natural Gas	Refined Petroleum	Electricity
Energy Sectors					
Coal	44.0	16.0	-30.0	23.0	-3.0
Coke & Coal Gas	52.0	24.0	-81.0	28.0	-6.0
Crude Petroleum & Natural Gas	211.0	27.0	-34.0	90.0	100.0
Refined Petroleum	161.0	26.0	-558.0	460.0	132.0
Electricity	277.0	9.0	-79.0	-4.0	5.0
Non-Energy Sectors					
Agriculture	40.0	10.0	-12.0	12.0	24.0
Metal Ores Mining	57.0	10.0	-32.0	76.0	7.0
Other Mining	88.0	14.0	-33.0	127.0	42.0
Food	64.0	10.0	-16.0	27.0	26.0
Textiles	79.0	14.0	-21.0	42.0	40.0
Apparel Products	74.0	13.0	-20.0	35.0	38.0
Sawmills	143.0	21.0	-30.0	50.0	34.0
Paper	122.0	19.0	-22.0	42.0	48.0
Chemicals	124.0	41.0	-49.0	43.0	50.0
Building Materials	162.0	37.0	-40.0	37.0	33.0
Primary Metals	125.0	68.0	-35.0	52.0	19.0
Metal Products	94.0	38.0	-28.0	48.0	17.0
Machinery	120.0	35.0	-25.0	45.0	36.0
Transport Equipment	126.0	34.0	-24.0	52.0	63.0
Electric Machinery	107.0	41.0	-26.0	51.0	35.0
Electronic Equipment	65.0	15.0	-23.0	30.0	16.0
Instruments & Meters	70.0	16.0	-21.0	35.0	25.0
Maintenance of Equipment	128.0	32.0	-25.0	68.0	47.0
Other Industries	136.0	46.0	-30.0	61.0	65.0
Construction	97.0	31.0	-29.0	59.0	31.0
Freight Transport	63.0	13.0	-73.0	102.0	32.0
Commerce	69.0	11.0	-25.0	37.0	28.0
Restaurants	61.0	4.0	-16.0	21.0	20.0
Passenger Transport	52.0	9.0	-81.0	21.0	28.0
Public Utilities	66.0	23.0	-28.0	31.0	9.0
Research Institutions	65.0	12.0	-19.0	39.0	16.0
Finance & Insurance	62.0	9.0	-19.0	43.0	9.0
Public Administration	86.0	12.0	-29.0	35.0	40.0

Source: SSDA calculations from "1995 China and Shanxi Province Input-Output Tables." *1995 Shanxi Statistical Yearbook.* China Statistics Press. *State Statistical Bureau of China (SSB). 1995. China Input-Output Table, 1995 (English Edition).* Beijing: China Statistical Press.

Fourth, the two regions are very similar in terms of crude oil & natural gas inputs. Just as is the case with energy direct inputs, Shanxi Province has lower direct-and-indirect inputs for crude petroleum & natural gas than China. That difference is fairly negligible, about 21% (782-614/782) in the five energy sectors, but much larger, more than 90% (831-41/831), in the 28 nonenergy sectors. Among the energy sectors, the most impressive difference is the amount of refined petroleum direct-and-indirect input into the refined petroleum sector (460.0), and among the nonenergy sectors, coal direct-and-indirect input into the building materials sectors (162.0).

4.3.4 Importance of Production Technology Changes

Our examination of the energy impacts of final-demand shifts presents only part of the energy comparison between Shanxi Province and China. The other part is the changes in production technology, which may increase or decrease the amount of intermediate energy used to deliver one unit of final goods and services. Because final-demand shifts alone explain a small portion (12.2%) of the differences in energy-intensity levels between Shanxi Province and China, production-technology differences must be responsible for a significant proportion of the extra energy consumption in Shanxi Province (Table 4.11).

As we just noted, differences in the mix final customers consume of goods and services contributed 12.2% of the differences between energy consumption in Shanxi Province and China. Only an additional 0.5 percent could be explained by the differences in the importance of the final-demand sectors between Shanxi Province and China, making a total of 12.8%. The two effects influence energy consumption because of the fact that there are variations among the 33 product sectors in terms of the amount of energy they consume to satisfy one unit of final demand. However, it is apparent that there must be another factor that also contributes to the differences in the total energy consumption between China and Shanxi Province. The remaining differences can be explained by the variations in production-technology levels of Shanxi Province and China.

4.3.5 Factors Influencing Production Technologies

There are five overlapping factors that could cause the actual technology of a given industry to be different. First, the types and quality of goods and services that are produced could be different. As an example, a shift from less to more energy-intensive products will lead to a decline in energy direct-and-indirect-input coefficients. Second, differences in production facilities, such as the introduction of a new facility line, a modification of existing facilities, or retirement of obsolete equipment, in one region could have profound effects on its energy consumption compared to another region. Third, changes in management practices and operations of production facilities, could improve energy direct-and-indirect-input coefficients, through such practices as improving operations of energy-intensive equipment and through better energy housekeeping. Fourth, differences in the quality of inputs could make a significant difference in energy consumption. Energy can be saved, by matching coal quality to the input specifications of a furnace or boiler or by switching from low-quality coal to high-quality coal. Fifth, changes in capacity utilization and scale of production could be of significance. We can divide input uses in production activities into two categories: those that vary with level of output and those that remain relatively constant when output changes. Within the limit of capacity, higher capacity utilization or output level reduces fixed inputs per unit of output, therefore reducing input requirements. An obvious example is the transportation sector. As more passengers (freight) travel in a given vehicle, fuel consumption per passenger (or unit of freight), generally, decreases (Holdren 1992).

Given these five factors, why is there such a significant difference in the technology-input structures between Shanxi Province and China? The answer at least partially lies in the degree to which China's policies interacted with the realities of Shanxi Province's macroeconomic needs and situations. The first has to do with Shanxi Province's role as being the largest energy producer in China, producing over 25% of the coal in China, and as such, there are few incentives to save energy, while at the same time, there is a tendency to adopt energy-intensive technologies (Li, Johnson, Changyi, Taylor, Zhiping, and Zhongxiao 1995).

In addition, during the 1980s, a severe energy shortage in China led to a major shift in energy policy, from a policy of complete devotion to increasing supply to one of placing equal emphasis on supply expansion and energy conservation, with priority given to conservation efforts in the short run (Tomitate 1989; Smil 1988). Starting in 1981, energy-conservation targets were incorporated into the Five-Year Social Economic Development Plan as well as into annual plans. They included targets for the amount of energy to be saved, limits on energy consumption in the production sectors, and energy-saving targets (Wang 1990). Overall, the energy-conservation program has been highly successful and has resulted in large energy savings (World Bank 1985). In some cases, however, the energy-conservation programs and plans were not as successful. They may not necessarily have been implemented at all, or as originally designed, at the local and enterprise level. This was especially true in Shanxi Province, where the availability of abundant and cheap coal, and the Province's reliance on coal for a majority of their industrial processes, made it much easier for the decision makers to ignore the directives of the national government. This is an illustration of the Chinese saying, "where the mountain is high, the emperor is far away," meaning that the central government only has limited control over local affairs.

Improvement in energy efficiency was also a by-product of China's economic reform and reflected the improvement in macroeconomic performance in the 1980s. Between 1949 and 1977, China's economic-development strategy followed the Soviet growth model, which emphasized high output growth and development of heavy industry. This strategy did result in high output growth, but at very high resource costs. The growth achieved was accompanied by great waste and inefficiencies, so that more and more investment and resources were required to attain a given increase in national income. Li, Jorgenson, Zheng, and Kuroda (1993) found that all the output growth in China from 1953 to 1978 came from increases in factor inputs; factor productivity actually went down during this period. Chen and Wang (1988) and Perkins (1986) reach a similar conclusion. In addition, managers of firms in this overly centralized economic system ignored supply and demand conditions and failed to produce what was needed. Some goods were overproduced and stockpiled, while others, especially consumer goods, were in chronic shortage. (Naughton 1987)

Since 1978, China has initiated a series of reform measures in the hopes that such measures would raise economic efficiency and the rate of productivity growth of its economy (Barnett and Clough 1986; Lampton 1987; Reynolds 1988; World Bank 1985, 1990). The reform started in rural areas in 1978 with the piecemeal dissolution of collective agriculture through the introduction of the household-land-contract or agricultural responsibility system. It penetrated into the urban economy in the mid-1980s in three basic forms: (1) greater decision-making autonomy for enterprises, in production and, to a lesser extent, in investment; (2) reinstitution of financial incentives for enterprises and individuals; and (3) expansion of the role of markets in the allocation of industrial goods and a corresponding reduction in the role of planning and administrative allocation (Byrd 1991, 1992). The basic objective of the reform was to make China's economy more market-oriented, that is, to shift from a centrally planned economy, in which planning and administrative directives guided the allocation of resources, to an eclectic, market-oriented, socialist commodity economy, in which resource allocation was determined largely by interactions in the market among autonomous, competitive, and profit-oriented economic agents. Although, Shanxi Province benefited from the above measures to increase output and decrease energy intensity, their starting energy intensity was so much greater than China as a whole, that the reforms merely kept Shanxi Province's energy-intensity levels at a much greater level than China's. In Shanxi Province as a whole, the level declined by 1998 to about the same level that China was at in 1978. Furthermore, because of Shanxi Province's position as the leading producer of energy, particularly coal, energy-intensive industries, such as coal mining and coke manufacturing, continue to play a significant role in the output levels of Shanxi Province.

Finally, improvements in energy efficiency were an enterprise's rational response to energy-price increases in China in the 1980s. For years, energy had been under-priced in China, partially in order to promote industrial development. In the early 1980s, coal prices, for example, were about 60% of the long-run marginal cost of coal production (World Bank 1985). The greatly under-priced energy provided no incentives for energy-efficiency improvements. Energy was consumed as if its value to the country was much lower than it really was. Processes were designed, machinery and appliances built, and buildings constructed that used more energy than was justified, considering its real value in other uses. Low prices also caused people to operate those facilities in ways that used more energy than they would have if managers had to account for energy's true value (World Bank 1985).

This problem of irrational energy pricing was mitigated, to some degree, by price reforms in the energy sector, mainly in the coal industry, in the early 1980s (Byrd 1987). Altogether in the 1980s the government raised the price of coal by 10 to 25%, and more fundamentally, they introduced a dual price system into the energy sector in 1983-1984 (Byrd 1987). Under this system, goods were exchanged at two different prices: a state-set price, for the amount produced under central planning, and a higher free-market price, for above-plan output. The State also removed price controls on locally produced coal, which accounted for an increasing share of total coal production (Byrd 1987; World Bank 1985). Although China's energy-price increases in the 1980s provided some incentives for energy-efficiency improvements, these incentives remained low because energy prices remained low despite the price hikes. Most fuels continued to be allocated through state planning, and the share of market-allocated energy was too small to affect the overall energy price structure. As such, due to low energy prices, energy expenditures made up a very small share of the total production costs; therefore, managers in most enterprises did not view cutting energy costs as a top priority. This situation was exasperated in Shanxi Province, because of the reasons mentioned in previous chapters concerning Shanxi Province's heavy reliance on coal as an energy input; its position as the primary coal producer in the country; and the fact that the fastest-growing production sectors were also some of the most energy-intensive sectors.

4.4 Conclusion

In this chapter, we conducted an in-depth analysis for Shanxi Province's energy-intensity change between 1992 and 1999 and showed the specific factors behind the energy-intensity difference between Shanxi Province and China. In the first part about the temporal energy-intensity changes in Shanxi Province, we found that in most of the sectors, the direct inputs of energy resources were decreasing during this period, but not for electricity, steam, & hot water. Furthermore, the total (direct-and-indirect) energy-input coefficient shows that all the sectors in both regions consumed less energy in 1999 than in 1992, in order to produce one unit of GDP. We show that petroleum refineries and cokemaking are the major contributors to the coal-intensity increase, and we conclude that technology improvement did help to reduce the use of most energy products significantly. The consumption of electricity, steam, & hot water is an exception. We believe that there is still room for technical improvements in efficiency in the use of coal and in the use of other energy resources.

In the second part, we showed with the use of SSDA that the differences in direct-and-indirect energy, especially coal, inputs, in the 33 sectors under observation was key to explaining the energy-intensity differences between Shanxi Province and China. Shifts in final demand contributed only 12.5% to the overall difference, while a negligible percentage can be attributed to the fact that composition in these five sectors was different between Shanxi Province and China. We also discuss some of the possible reasons behind these energy-intensity differentials, almost all of which can explain the reductions in energy-intensity levels in both China and Shanxi Province, while also giving a clue as to why there is such a difference between the two regions. We especially believe that energy-conservation programs, the energy-pricing system, and the macroeconomic performance played important roles.

Cokemaking is obviously one of the sectors that is contributing to the vast differences in energy use between China and Shanxi Province. Given the survey data described in Chapter 2, we suggest that analysts could conduct more detailed analyses in the future by taking the energy and other input information for different types of coke ovens and incorporating them into the SSDA analyses.

Appendix A-4. A Note on Data

The 1992 Shanxi input-output table has 33 sectors, while the 1999 table has 49 sectors. Before doing the SDA calculation, we converted the 1999 table from 49 sectors to 33 sectors. First, we aggregated the 49 sectors into 32 sectors, in which the petroleum refineries and cokemaking, gas, & coal products is listed as one single sector. Second, we split it into a petroleum-refinery sector and a cokemaking, gas, & coal products sector. To split the sector, we used 1997 national input coefficients for 124 sectors in which "Cokemaking" and "Petroleum refineries" are separate, as a reference. After aggregating the 124-sector table to 33 sectors, we calculated the proportion of the input coefficients in the rows of petroleum and cokemaking; use the proportion to split the petroleum refineries and cokemaking, gas, and coal products sector into two sectors, Cokemaking" and "Petroleum refineries." We made all the calculations using the converted 33-sector Shanxi input-output table.

In addition, in both the 1992 and 1999 Shanxi input-output tables, many elements are zero, but should not be. These numbers, therefore, must not have been available. Especially, all the 1999 export data are not available. For calculation, the editor of the table (Shanxi Statistics Bureau) used zeroes. Given that Shanxi is a major energy-product export province, our SDA calculations are not as accurate as we would like them to be, but are the best we could obtain. Moreover, when we searched for other data, such as energy-consumption distributions, we could not find appropriate data. Data are becoming increasingly available in China as well as in Shanxi Province, and we look forward to conducting additional analyses in the future with improved data.

References

Barnett, A. Doak, and Ralph N. Clough, eds. 1986. *Modernizing China: Post-Mao Reform Development.* Boulder: Westview Press.

Byrd, William. 1987. The Impact of the Two-Tier Plan/Market System in Chinese Industry. *Journal of Comparative Economics*, **11**(3) (March): 295-308.

Byrd, William. 1991. *The Market Mechanism and Economic Reforms in China.* Armonk, NY: M.E. Sharpe.

Byrd, William. 1992. *Chinese Industrial Firms under Reform.* New York: Oxford University Press.

Carter, Anne. P. 1970. *Structural Change in the American Economy.* Cambridge: Harvard University Press.

Chen, Yizi and Wang Xiaoqiang. 1988. Reform: Results and Lessons from the 1985 CESRRI Survey. In *Chinese Economic Reform: How Far, How Fast?* edited by Bruce Reynolds. Boston: Academic Press: 172-188

Hannon, Bruce. 1983. *Analysis of the Energy Cost of Economic Activities: 1963 to 1980.* ERG Document 316. Urbana: University of Illinois at Urbana-Champaign, Energy Research Group.

Holdren, John. 1992. "Prologue: The Transition to Costlier Energy." In *Energy Efficiency and Human Activity: Past Trends, Future Prospects.* Lee Schipper and Stephen Meyer, Eds. Cambridge: Cambridge University Press: 1-51.

Lampton, David M., ed. 1987. *Policy Implementation in Post-Mao China.* Los Angeles: University of California Press.

Leontief, Wassily. 1936. Quantitative Input and Output Relations in the Economic System of the United States. *Review of Economics and Statistics,* **11**(2): 105-125. February

Leontief, Wassily. 1985. The Choice of Technology. *Scientific American*, **252**(6): 37-45. June.

Li, Jingwen, Dale W. Jorgenson, Youjing Zheng, and Masahiro Kuroda. 1993. *Productivity and Economic Growth in China, the USA, and Japan* (translated to English from Chinese). Beijing: China Social Services Press.

Lin, Xiannuan. 1996. *China's Energy Strategy: Economic Structure, Technological Choices, and Energy Consumption.* Westport, CT: Praeger.

Lin, Xiannuan, and Karen R. Polenske. 1995. Input-Output Anatomy of China's Use Changes in the 1980s. *Economic Systems Research.* 7(1): 67-84.

Naughton, Barry. 1987. The Decline of Central Control over Investment in Post-Mao China. In *Policy Implementation in Post-Mao China*, edited by David M. Lampton. Los Angeles: University of California Press: 51-80.

Office of Technology Assessment (OTA). 1991. *Energy in Developing Countries*. OTA-E-486. Washington DC: U.S. Government Printing Office.

Perkins, Dwight H. 1986. The Prospects of China's Economic Reforms. In *Modernizing China: Post-Mao Reform and Development*, edited by A. Doak Barnett and Ralph N. Clough. Boulder: Westview Press: 39-62.

Ploger, Ellen. 1985. The Effects of Structural Changes on Danish Energy Consumption. In *Input-Output Modeling*, edited by A. Smyshlyaev. New York: Springer-Verlag: 211-220.

Polenske, Karen R., and Stephen F. Fournier. 1993. Conceptual Input-Output Accounting and Modeling Framework. In *Spreadsheet Models for Urban and Regional Analysis*, edited by Richard E. Klosterman and Richard K. Brail. New Brunswick, NJ: Center for Urban Policy Research: 205-228.

Proops, John L.R. 1984. Modeling of the Energy-Output Ratio. *Energy Economics* **6**(1): 47-51.

Reardon, William A. 1976. *An Input-Output Analysis of Energy Use Changes from 1947 to 1958, 1958 to 1963, and 1963 to 1967*. A Report Prepared for the Electric Power Research Institute. Richland, WA: Battle Pacific Northwest Laboratory.

Reynolds, Bruce L. 1988. Trade, Employment, and Inequality in Post-Reform China. In *Chinese Economic Reform: How Far, How Fast?*, edited by Bruce Reynolds. Boston: Academic Press: 189-199.

Rose, Adam, and Stephen D. Casler. 1996. Input-Output Structural Decomposition Analysis: A Critical Appraisal. *Economic Systems Research* 8(1): 33-62.

Shanxi Province, 1999 Shanxi Input-Output Table. *1999 Shanxi Statistical Yearbook.* Beijing: China Statistics Press.

Shanxi Province. 1997. *Shanxi Province Statistical Yearbook, 1997*. Beijing: China Statistics Press.

Shanxi Province. 1995. 1995 Shanxi Province Input-Output Table. *1995 Shanxi Statistical Yearbook.* Beijing: China Statistics Press.

Shanxi Province. 1992. 1992 Shanxi Province Input-Output Table. *1992 Shanxi Statistical Yearbook.* Beijing: China Statistics Press.

Sinton, Jonathan. 1996, *China Energy Databook*. Lawrence Berkeley National Laboratory, September 1996.

Smil, Vaclav. 1988. *Energy in China's Modernization*. New York: M.E. Sharpe.

SSB (State Statistical Bureau of China). 1997a. *China Statistical Yearbook, 1997* (English Edition). Beijing: China Statistical Publishing House.

SSB (State Statistical Bureau of China). 1997b. *Gross Domestic Product, 1952-1995* (English Edition). Beijing: China Statistical Publishing House.

SSB. (State Statistical Bureau of China). 1996. *China Energy Annual Review*. Beijing: China Statistical Publishing House.

SSB (State Statistical Bureau of China). 1995. 1995 China Input-Output Table. Beijing: China Statistical Publishing House.

SSB. (State Statistical Bureau of China). 1994. *China Energy Annual Review*. Beijing: China Statistical Publishing House.

SSB (State Statistical Bureau of China). 1993. *1991 China Energy Statistical Yearbook* (English Edition). Beijing: China Statistical Publishing House.

SSB (State Statistical Bureau of China). 1992. *China Input-Output Table, 1992* (English Edition). Beijing: China Statistical Publishing House.

SSB (State Statistical Bureau of China). 1991. China Provincial Statistics, 1949-1989. (English Edition) Beijing: China Statistical Publishing House.

Strout, Alan M. 1966. *Technological Change and the United States Energy Consumption, 1939-1954*. Ph.D. Dissertation. Chicago IL: University of Chicago.

Tomitate, Takao. 1989. Economic Development and Energy Problems in China. *Energy in Japan.* **95**(3): 20 (March).

Wang, Haibo. 1990. *On Industrial Economic Efficiency* (translated to English from Chinese). Beijing: Economic Management Press.

World Bank. 1990. *China: Socialist Economic Development*. Washington, DC: World Bank.

World Bank. 1985. *China: Long-Term Development Issues and Options. Annex 3: The Energy Sector.* Washington DC: World Bank.

---

[1] All authors were members of the multiregional planning (MRP) research group at the Massachusetts Institute of Technology, Department of Urban Studies and Planning, when the main part of this research was conducted. Currently, Mahdavi is Strategy Consultant, Accenture, Minneapolis, MN; Guo Wei SQA (Software Quality Assurance) Engineer, Bit9, Cambridge, MA, and Polenske is Professor of Regional Political Economy and Planning; Head, China Cokemaking Team, Department of Urban Studies and Planning, Massachusetts Institute of Technology, USA.

CHAPTER 5

ENERGY EFFICIENCY AND PROFITABILITY DIFFERENCES: STATE-OWNED VERSUS TOWNSHIP-AND-VILLAGE ENTERPRISES

Ali SHIRVANI-MAHDAVI[1]

5.0 Introduction

There is a paradox between the energy efficiency and profitability in China's township-and-village enterprises (TVEs) and non-township-and-village enterprises (NonTVEs), which consist mostly of State-Owned Enterprises (SOEs), where we define profitability as profits per 100 RMB of capital. TVEs have experienced dramatic growth since they first appeared in 1978. Originally, TVEs were conceived as enterprises to be run by townships or villages, to promote economic growth, absorb the surplus rural labor force, and to discourage excessive urban migration. Today, TVEs include a wide range of cooperative and individual enterprises often run by members of the rural population. Interestingly, although TVEs are generally more profitable than SOEs, TVEs are far less energy efficient. To tackle this paradox, we first give an overview of the differences between SOEs and TVEs. Then, we examine the underlying energy and profitability differences between these firm-types by using a structural decomposition analysis. Finally, we give a theoretical basis for the observed paradox, focusing on property-rights issues.

5.0.1 State-Owned Enterprises versus Township-and-Village Enterprises

At the end of this chapter, we describe in detail how the ownership of SOEs and TVEs has changed over time. Here, we provide only an overview of the present-day differences. State-owned enterprises are owned and run by provincial or national governments. TVEs, on the other hand, are collectively owned and run by township or village governments.

Karen R. Polenske (ed.), The Technology-Energy-Environmental-Health (TEEH) Chain in China: A Case Study of Cokemaking, 71–89.

There are at least four key differences between SOEs and TVEs. First, TVEs face stricter budget constraints than SOEs (Steinfeld 1998; Perotti, Sun, and Xu 1999). Further, unlike SOEs, TVEs do not receive easy "policy loans" from the central banking system, because all TVEs are historically institutionalized as part of the traditional rural sector, whereas the banking system is part of the modern urban system (Steinfeld 1998). As a consequence, State banks have typically followed commercial principles in making loans to TVEs. Often, they ask Township-and-Village Governments (TVGs), which oversee TVEs, to act as guarantors of investment loans (Perotti, Sun, and Xu 1999).

The second difference between SOEs and TVEs is that historically, SOEs have had many objectives other than production and profit seeking. Among them are political support of the government, expansion of employment, and provision of various social services and securities, such as housing, education, health insurance, unemployment payments, and pensions (Perotti, Sun, and Xu 1999). As such, this economic burden of providing a large set of public goods to its employees has severely lessened the profitability of SOEs. Bell, Khor, and Kochlar (1993) estimate that about 20 percent of employees in the SOE sector are redundant. According to Xiao (1991), 40 percent of the difference in profitability between SOEs and TVEs can be attributed to social provisions of the kind described above. In addition to the direct contribution, SOEs provide de facto unemployment insurance payments to their redundant employees, also referred to as on-the-job unemployment.

A third difference between the two sectors is the process of investment decision making in the SOE sector. According to Sun (1998: 89), "the process of investment decision making in the State sector is a distribution process of rights to possess and use certain scarce State assets, including budget funds, bank loans, land, quotas of power, oil, and other key materials." As such, the first intention of SOEs is to obtain and occupy as much investment and property from the distributive negotiation process as possible, so that they can reap future benefits and justify their power base (Sun 1998). The consequence then is that when trying to establish new investment projects, the decision-makers do not care much about whether or not the project would be profitable in the long run, although that is changing dramatically as the reform process moves forward. However, for a long time, this form of investment-expansion drive, combined with the persistent soft-budget constraint, led to inefficient investment projects (Zou and Sun 1996).

Finally, despite their overall superior profitability, TVEs, in general, happen to be less energy efficient than SOEs—a contradiction we discuss in detail in the next section.

5.0.2 Energy and Profitability

Our team first came across the paradox between energy efficiency and profitability early in our first surveys of the coke industry in Shanxi Province (1998-2000). We had conducted the SOE (which, again, comprise the main part of the NonTVE sector) and TVE coke plants surveys to help us determine the relationship between technology, energy, environment, and health in the coke industry in Shanxi Province (Chapter 2, Section 2).

Of the over 192 firms we surveyed in 1999, 67 percent of TVE cokemaking plants claimed to consume 1.39 metric tonnes or more of coal to produce one metric tonne of coke. In contrast, almost 70 percent of SOEs consumed 1.4 metric tonnes of coal or less to produce one tonne of coke. Also, a much larger proportion of TVEs reported net profits five years prior to our surveys than did the surveyed SOEs.

Using structural decomposition analysis (SDA), I answer two questions. First, would the same energy/economy characteristics hold true for other SOE and TVE sectors in Shanxi Province and China? Second, would differences in direct and indirect labor and other-material inputs partially offset the extra direct and indirect energy inputs that would then result in the superior profitability of TVEs? In the next section, I conduct an input-output analysis for Shanxi Province, and in the following section, I expand the analysis to China as a whole.

5.1 Direct and Indirect Input Differences Between SOEs and TVEs, Shanxi Province, 1995

There are two points to keep in mind when evaluating the following results. First, we chose the 29 sectors directly from the input-output tables that were constructed for Shanxi Province and China. Second, and very important for our discussion, is that all the values in the input-output tables that we used for this study are in monetary units. This is significant because it introduces the differences in input prices and labor costs between SOEs and TVEs, which we discuss in detail later. Figures 5.1, 5.2, 5.3, and 5.4 show the results of the input-output analysis for Shanxi Province. Figure 5.1 shows the difference in direct and indirect energy inputs, Figure 5.2 shows the direct and indirect labor inputs, and Figure 5.3 shows the comparison between the percentage of total output, total energy consumed, and total labor inputs between SOEs and TVEs in Shanxi Province for 1995. Finally Figure 5.4 shows the direct and indirect input of all other material inputs between SOEs and TVEs in Shanxi Province.

There are two observations that can be made from Figure 5.1. First, the monetary value of energy inputs is lower for SOEs than for TVEs in 28 out of 29 sectors in Shanxi Province, the only exception being the crude petroleum and natural gas sector. Overall, among these 28 sectors, directly and indirectly, TVEs spend between 33 and 50 percent more on energy to generate one RMB of output than SOEs. This is a significant difference in energy-input levels between the two sectors, and as such should have a profound effect on the economic performance of the two sectors. Second, there is a significant difference in the level of direct and indirect energy inputs in the 5 energy industries and the 24 non-energy industries, regardless of whether the inputs were consumed by SOEs or TVEs. This result is similar to those from previous studies (Shirvani-Mahdavi 1999) that show that intra-industry input levels are generally higher than other industry input levels. The results of this analysis show that the same holds true regardless of the sector under examination.

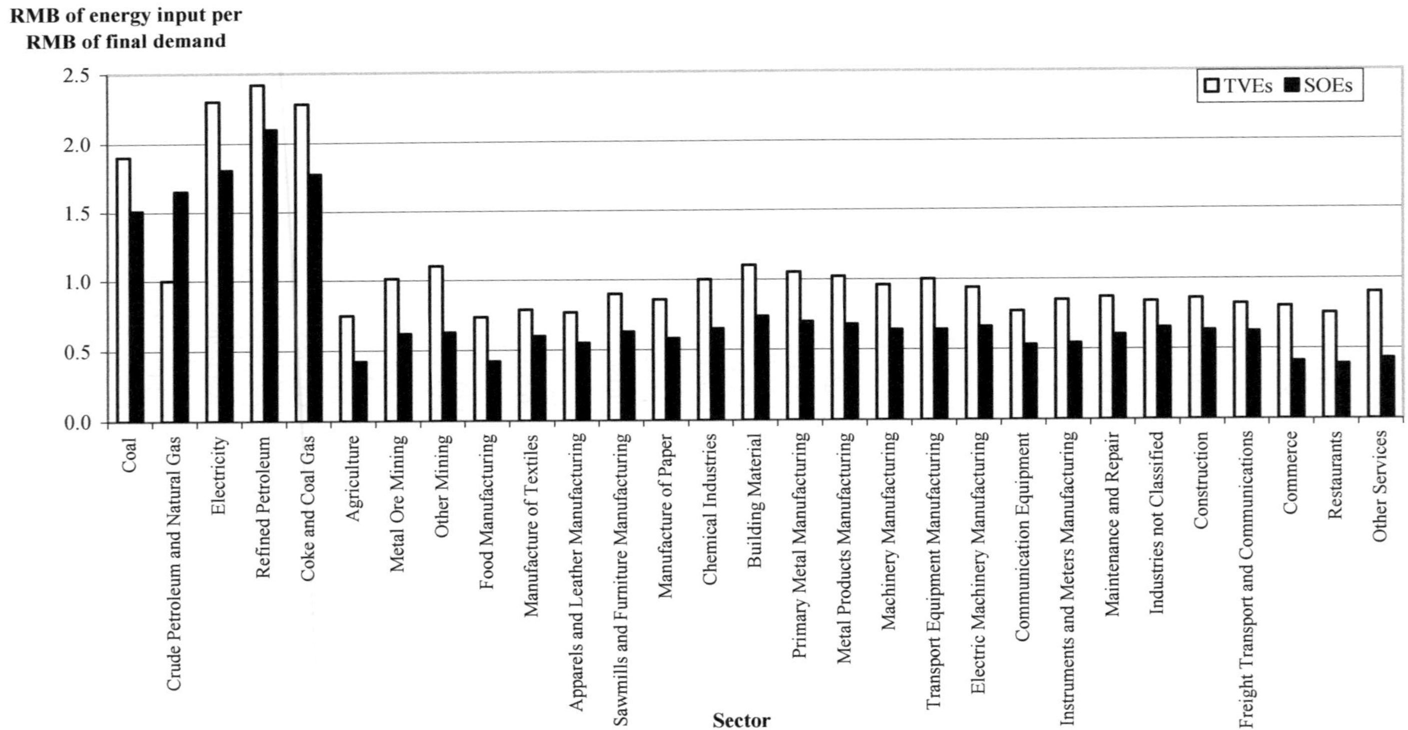

Note: SOEs (State-Owned Enterprises) include all NonTownship and Village Enterprises, but is comprised almost entirely of SOEs.
TVEs: Township-and-Village Enterprises.

Figure 5.1: Direct and Indirect Energy Input, SOEs and TVEs, Shanxi Province, 1995
(Direct and Indirect Input per RMB of Final Demand)

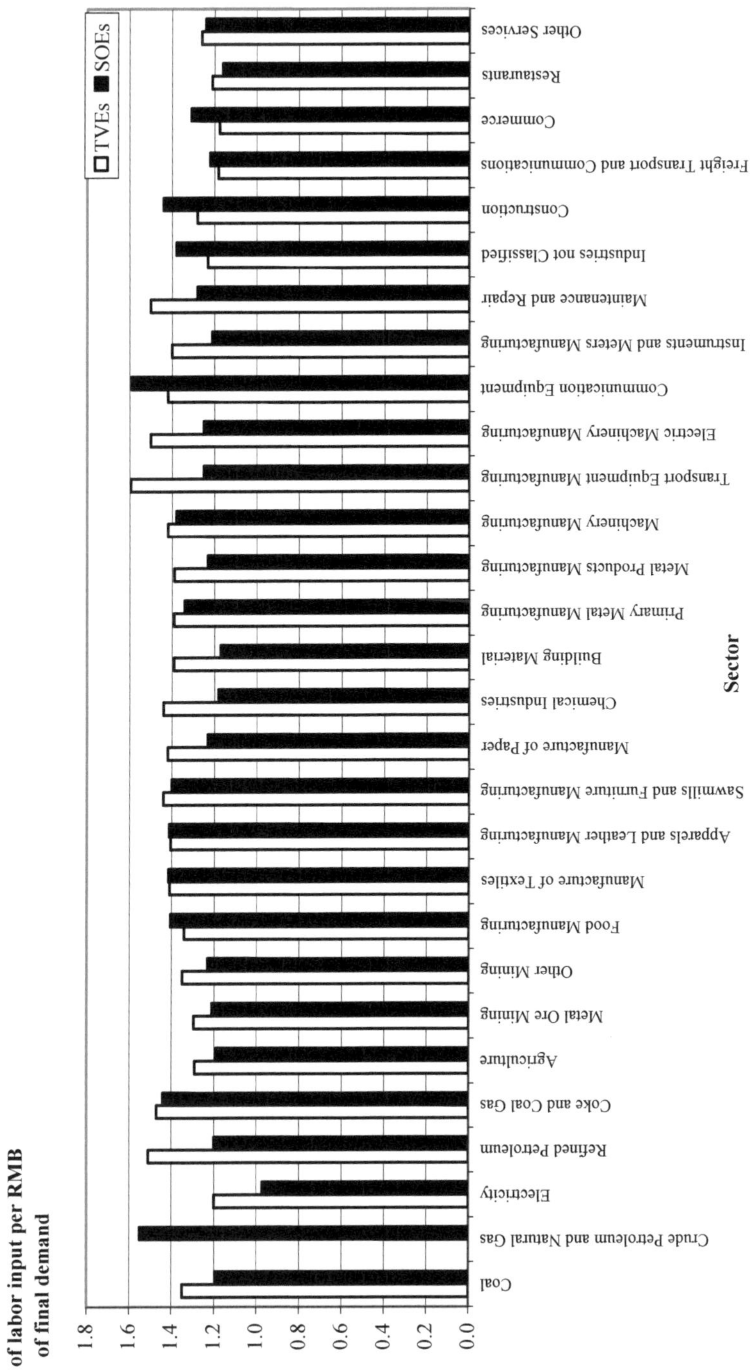

Note: For this study, SOEs (State-Owned Enterprises) include all NonTownship and Village Enterprises, but is comprised almost entirely of SOEs. Data on labor input into Crude Petroleum and Natural Gas of TVEs are not available.

Figure 5.2: Direct and Indirect Labor Input, SOEs and TVEs, Shanxi Province, 1995 (Direct and Indirect Input per RMB of Final Demand)

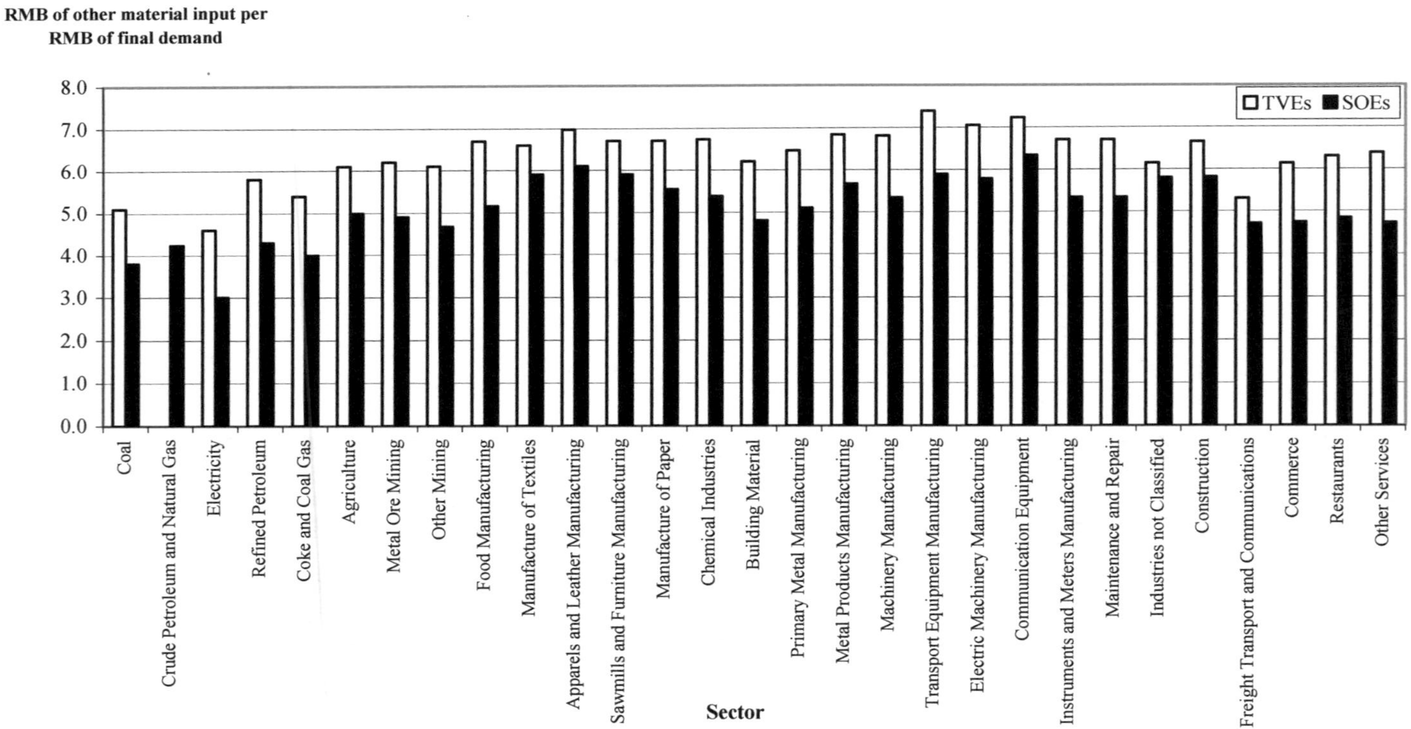

Note: For this study, SOEs (State-Owned Enterprises) include all NonTownship and Village Enterprises, but is comprised almost entirely of SOEs. Data on Crude Petroleum and Natural Gas of TVEs are not available.

Figure 5.3: Direct and Indirect Input of Other Material, SOEs and TVEs, Shanxi Province, 1995 (Direct and Indirect Input per RMB of Final Demand)

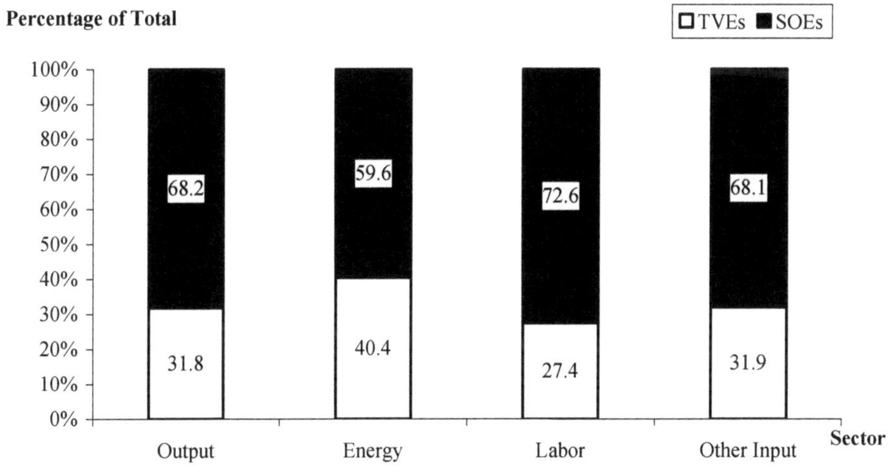

Note: For this study, SOEs (State-Owned Enterprises) include all NonTownship and Village Enterprises, but is comprised almost entirely of SOEs.

Figure 5.4: Percentage of Total Output, Energy Consumption, Labor Consumption, and Other Inputs, TVEs and SOEs, Shanxi Province, 1995

The next question is whether TVEs enjoy an advantage over SOEs in terms of labor input that makes up for the increased levels of energy input. Figure 5.3 shows the comparison of direct and indirect labor input, in monetary terms, between SOEs and TVEs. In this instance, no clear patterns stand out between the two sectors. In the five energy sectors, TVEs tend to have somewhat higher levels of labor input than SOEs, with the exception of the crude petroleum and natural gas sector, which we have already discussed. Among the remaining sectors, TVEs appear to have higher labor input levels than SOEs in the heavy industrial sectors (mining, chemical industries, and manufacturing sectors), while SOEs have greater labor input costs than TVEs in the commerce, construction, and communication sectors. Overall, however, no significant difference or pattern seems to occur in the direct and indirect labor input levels between the two sectors.

Our final analysis of the direct and indirect input levels is the monetary amount of other-material inputs that goes into producing one RMB of final demand. Figure 5.3 shows the results of this analysis for Shanxi Province. This chart mirrors the results of Figure 5.1, showing the same discrepancy in the levels of direct and indirect inputs between SOEs and TVEs across the sectors, the one exception being the crude petroleum and natural gas sector. Similar to the direct and indirect energy inputs, TVEs, on average, consume 15% additional monetary other-material inputs than SOEs, with input values ranging from 40 percent for electricity production to 5 percent for freight transport and communications of the total cost of production. Finally, Figure 5.4 shows the combined results of Figures 5.1, 5.2, and 5.3, for all 29 SOE and TVE sectors. This is a simpler representation of the differences in the levels of output, energy consumption, labor compensation, and other material inputs than the previous three figures, because it shows only the direct proportion of total inputs in each case, rather than the direct and indirect inputs shown in Figures 5.1 and 5.2. Despite the change in the level of analysis, the same trends hold true for energy consumption and labor compensation. In 1995, TVEs produced almost 31 percent of total output in monetary terms, but consumed over 40 percent of the energy, and spent only 27 percent on labor compensation, slightly less than the proportion of total output.

5.2 Direct and Indirect Inputs: SOEs and TVEs, China, 1995

We have just shown that almost all TVE sectors in Shanxi Province exhibited the same pattern of lower energy efficiency in 1995 that we observed in the TVE cokemaking sector. In this section, we conduct the same comparisons for the SOE and TVE sectors for China as a whole.

Figure 5.5 shows the results of the energy input-output analysis between SOEs and TVEs for China in 1995. There are a number of similarities between Figures 5.1 and 5.5. First, just as is the case in Shanxi Province, the direct and indirect monetary energy inputs are greater for TVEs than for SOEs, in 28 out of 29 sectors, including crude petroleum and natural gas. The only aberration is the coke and coal-gas sector, for which the SOEs have slightly greater energy-input levels than TVEs. In addition, similar to Shanxi Province, the largest direct and indirect inputs come from the sector itself, and this is the case for all five energy sectors. This is just as true for TVEs as it is for SOEs. Furthermore, similar to Shanxi Province, the sectors where the differences in energy inputs are the greatest are the heavy industry and manufacturing sectors, such as mining and manufacturing sectors.

There is however a profound difference between Shanxi Province and China, and that is for the overall levels of direct and indirect energy inputs. In Shanxi Province, the energy-input levels for TVEs range from 2.7 Renminbi (RMB) of energy input per RMB of final demand in the refined petroleum sector to 0.7 RMB of energy input for one RMB of final demand in the restaurant sector. In China, however, the range of energy inputs is 1.9 RMB per RMB of final demand for the TVE refined petroleum sector, to 0.22 RMB of energy input per RMB of final demand for restaurants.

Figure 5.6 shows the direct and indirect labor inputs for SOE and TVE sectors for China in 1995. Unlike energy, there are major differences between Shanxi Province and China, in terms of labor input for SOEs and TVEs. As is apparent from this chart, the direct and indirect monetary labor input is greater for the SOE sector than for the TVE sector for 28 of 29 sectors, the exception being commerce, where there is a negligible difference in labor input between SOEs and TVEs. This is a clear indication of where TVEs have the advantage over SOEs in terms of profitability. Again, we stress that given the monetary nature of the analysis, the results here do not indicate the productivity of the perspective labor forces, and the difference in input could entirely be due to wage levels and other benefits that SOE employees enjoy as compared to TVE employees.

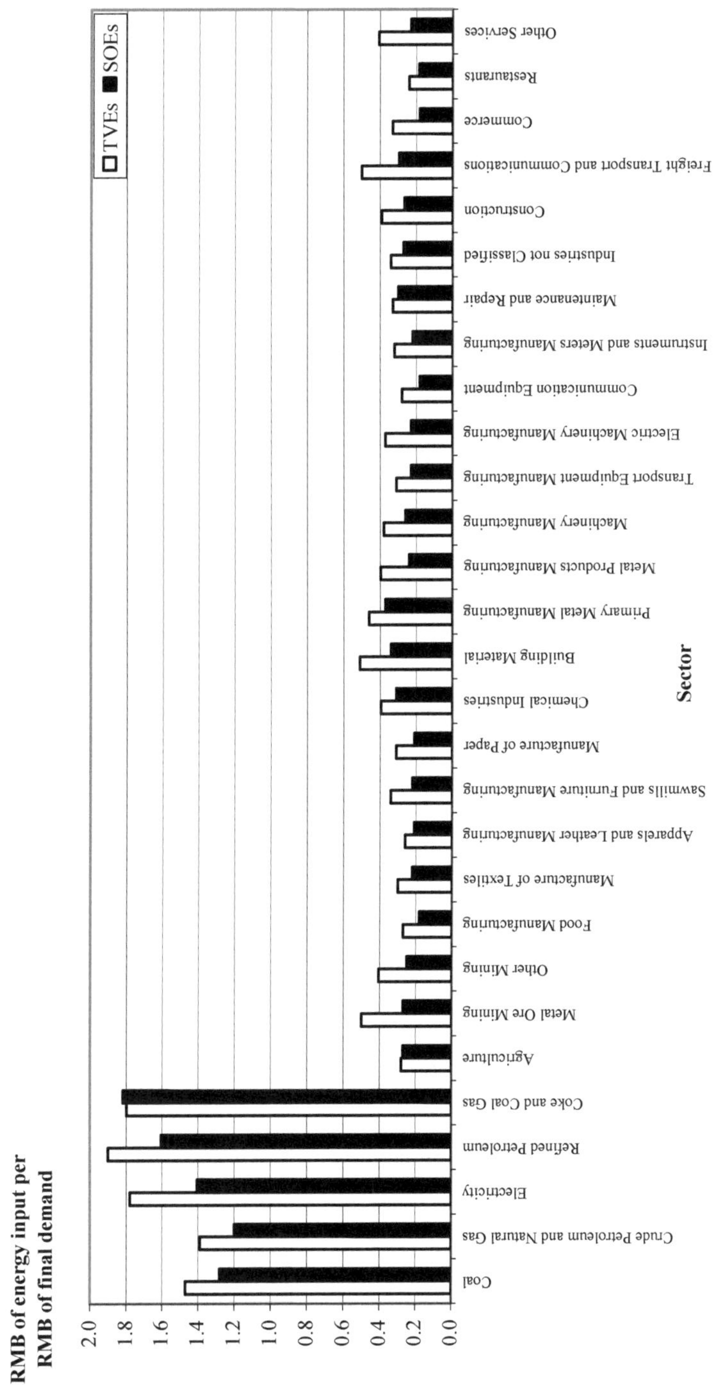

Note: For this study, SOEs (State-Owned Enterprises) include all NonTownship and Village Enterprises, but is comprised almost entirely of SOEs.

Figure 5.5: Direct and Indirect Energy Input, SOEs and TVEs, China, 1995
(Direct and Indirect Input per RMB of Final Demand)

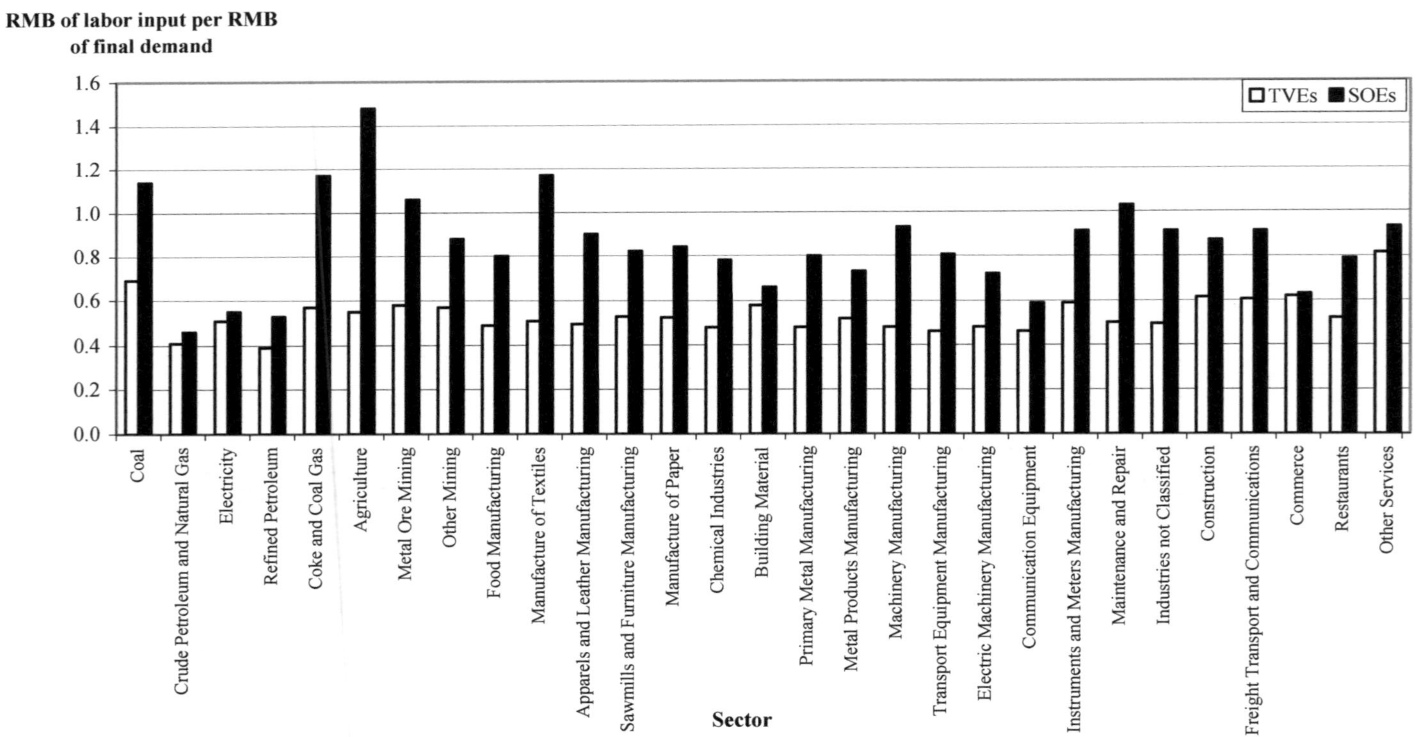

Note: For this study, SOEs (State-Owned Enterprises) include all NonTownship and Village Enterprises, but is comprised almost entirely of SOEs.

Figure 5.6: Direct and Indirect Labor Input, SOEs and TVEs, China, 1995
(Direct and Indirect Input per RMB of Final Demand)

Similar to Shanxi Province, the final part of this analysis entails the direct and indirect input requirements for other materials by the SOEs and TVEs. Figure 5.7 shows the results for China in 1995. Similar to Shanxi Province, most TVE sectors have larger direct and indirect other-material inputs than SOEs. However, unlike Shanxi Province, the difference is not across the board, and it is not as large. Furthermore, similar to the direct and indirect energy inputs, the overall levels of other-material inputs is much larger for Shanxi Province than China, ranging from 7.2 RMB of other-material inputs per RMB of final demand for TVE transport-equipment manufacturing to 4.3 RMB of inputs per RMB of final demand for TVE electricity production in Shanxi Province. In China, on the other hand, the range is from 5.2 RMB of inputs per RMB of final demand for the SOE construction sector to 2.5 RMB inputs per RMB of final demand for TVE refined petroleum processing. It appears then, that, in general, the level of direct and indirect material inputs is larger for all SOE and TVE sectors in Shanxi Province than in China.

Figure 5.8 shows a summary table of the direct input requirements for energy, labor, and other materials for China in 1995. This figure shows the same general trends as Figure 5.4 did for Shanxi Province, that is, the TVEs' share of energy consumption is larger than their share of total output, while their share of compensation to labor is lower than their share of output. However, there are also differences. In China, SOEs' share of labor compensation is greater than their share of total output. In fact, the 14% difference is very large, and it plays a major role in the profitability of the SOE sector in China. Second, in China, TVEs extra consumption of material inputs is split more evenly between energy inputs and other material inputs, whereas in Shanxi Province, the majority of extra inputs is in the form of extra energy consumption, which is primarily due to Shanxi Province's access to a large amount of inexpensive fuels, namely coal.

5.3 Profitability Differences Between SOEs and TVEs in China, 1995

In the first part of this paper, we used the profitability levels of SOE and TVE coke plants in Shanxi Province to claim that the TVE cokemaking sector is more profitable than the SOE cokemaking sector. Here, we expand the number of criteria for profitability, and we extend the region of analysis and sectors to include all of China and all sectors, respectively. Our analysis shows that TVEs are indeed more profitable than SOEs, and that the paradox of energy inefficiency and profitability of TVEs over SOEs can be expanded from the cokemaking sector in Shanxi Province to include SOE and TVE sectors as a whole in China. Table 5.1 summarizes some relevant economic data for the SOE and TVE sectors for China for the period 1979 to 1995. In this case, I calculated total factor productivity (TFP) as a residual after subtracting from output growth a weighted average of the growth rates of labor and capital inputs. The weights I use are 0.7 for labor and 0.3 for capital, corresponding crudely to rough worldwide consensus that labor's share of income is between 0.66 and 0.75 (Weitzmann and Xu 1994).

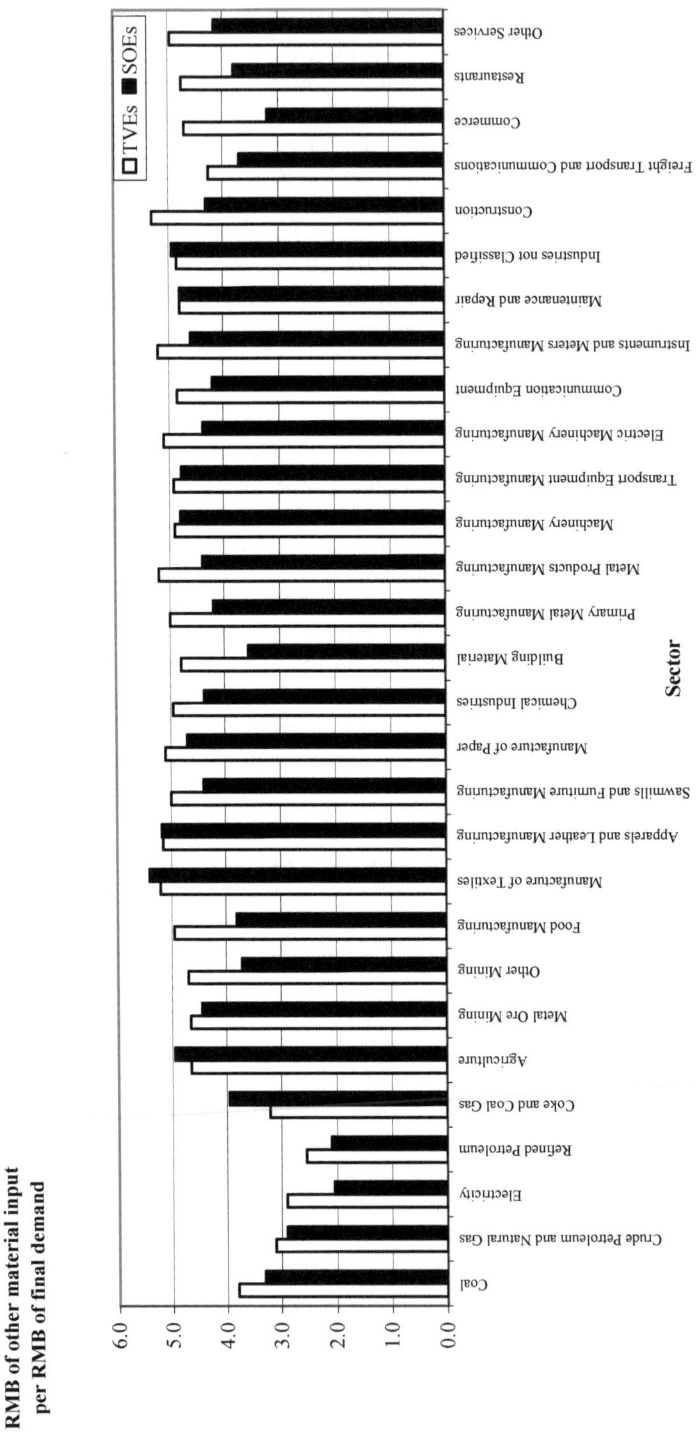

Note: For this study, SOEs (State-Owned Enterprises) include all NonTownship and Village Enterprises, but is comprised almost entirely of SOEs.

Figure 5.7: Direct and Indirect Input of Other Material, SOEs and TVEs, China, 1995 (Direct and Indirect Input per RMB of Final Demand)

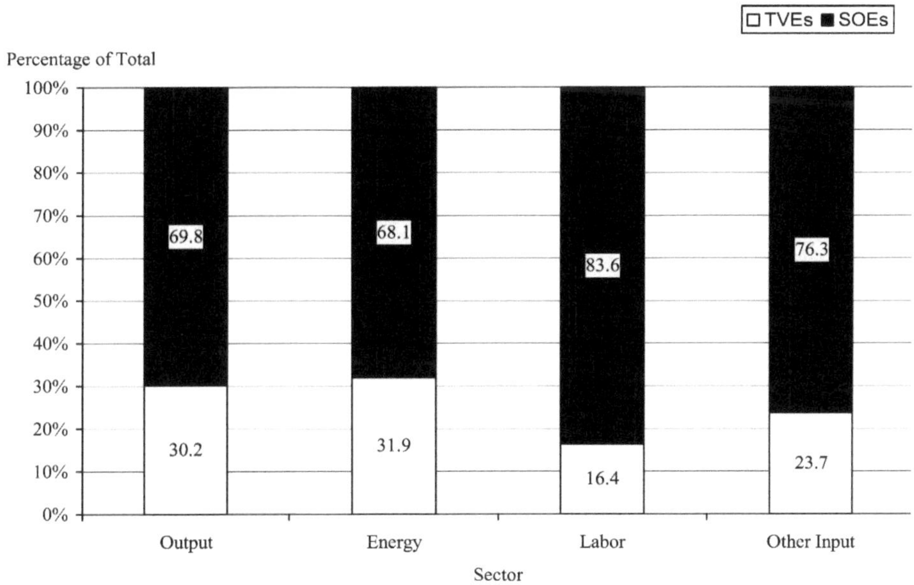

Note: For this study, SOEs (State-Owned Enterprises) include all NonTownship and Village Enterprises, but is comprised almost entirely of SOEs.

Figure 5.8: Percentage of Total Output, Energy Consumption, Labor Consumption, and Other Inputs, SOEs and TVEs, China, 1995

Together, Tables 5.1 and 5.2 reveal several important insights. First, total factor productivity has grown approximately three times faster for TVEs than for the corresponding SOEs, and TVEs enjoy greater capital efficiency than SOEs. In fact, the tables show that in almost every respect, TVEs are more capital efficient than SOEs. For example, TVEs' profits per 100 RMB of capita are over 2.5 times those of SOEs (16.1 RMB versus 6.4). Similarly, the output value per 100 RMB of capital is almost twice as much, and value added per 100 RMB of capital is 1.5 times as much as SOEs. Furthermore, capital and taxes per employee is 66 percent greater for TVEs than SOEs, value added per employee is almost twice as great, while capital expenditures per employee are almost three times as much. In short, TVEs generated more taxes, profits, and output per RMB of capital expenditures and number of employees than SOEs in China in 1995.

The same patterns hold true when we expand the level of analysis to include all SOE and TVE sectors, such as agriculture and commerce. Table 5.3 shows the results of this analysis. Three phenomena stand out from this table. First, similar to industrial TVEs, all TVE sectors are more capital efficient than SOEs. As an example, profits per 100 RMB of capital are almost five times as great in the TVE sector as in the SOE sector, whereas the following are almost twice as much: (1) profits per 100 RMB of sales revenue, (2) sales revenue per 100 RMB of capital, and (3) profits and taxes per 100 RMB of capital. In 1995, TVEs generated a great deal more profits, revenue, and taxes per RMB of capital investment than SOEs.

Table 5.1: Growth and Efficiency in SOE and TVE Sectors, 1979-1995

Growth Rate	TVEs (%)	SOEs (%)	Ratio of TVEs to SOEs
Output, 1979-1995	25.3	8.4	3.0
Capital, 1979-1995	16.5	7.8	2.1
Labor, 1979-1995	11.9	3.0	4.0
Total Factor Productivity, 1979-1995	12.0	4.0	3.0

Source: China Statistical Yearbook, various years;
Weitzman and Xu. (1994).

Note: For this study, SOEs (State-Owned Enterprises) include all NonTownship and Village Enterprises, but is comprised almost entirely of SOEs.

Table 5.2: Selected Profitability Indicators for Industrial SOEs and TVEs, China, 1995

Indicator	TVEs	SOEs	Ratio of TVEs to SOEs
Taxes (billion RMB)	187.56	412.78	0.45
Profits (billion RMB)	366.03	463.49	0.78
Compensation of Employees (billion RMB)	264.55	534.81	0.49
Value Added (billion RMB)	887.14	1748.18	0.50
Output Value (billion RMB)	3,621.93	5,567.57	0.65
Labor Force (million persons)	71.24	75.21	0.94
Total Capital (billion RMB)	2,280.24	7,271.63	0.31
of which: fixed assets (billion RMB)	1,097.53	4,225.00	0.26
Profits and Taxes per 100 RMB Capital (RMB)	24.28	17.05	1.42
Profits per 100 RMB Capital (RMB)	16.05	6.37	2.52
Output Value per 100 RMB Capital (RMB)	158.84	76.57	2.07
Value Added per 100 RMB Capital (RMB)	38.91	24.04	1.62
Profits and Taxes per 100 RMB Output (RMB)	15.30	15.70	0.97
Profits and Taxes per Employee (RMB)	7,771.00	11,651.00	0.67
Compensation per Employee (RMB)	3713.00	7,111.00	0.52
Value Added per Employee (RMB)	12,453.00	23,244.00	0.54
Capital per Employee (RMB)	32,008.00	96,684.00	0.33

Source: Chen et al. (1999).

Note: RMB: Renminbi, the Chinese currency, equivalent to $0.125.
For this study, SOEs (State-Owned Enterprises) include all NonTownship and Village Enterprises, but is comprised almost entirely of SOEs.

Table 5.3: Selected Profitability Indicators for All SOEs and TVEs, China, 1995

	TVEs	SOEs	Ratio of TVEs to SOEs
Total Profit and Taxes (billion RMB)	314.72	552.19	0.57
of which: profits (billion RMB)	196.93	145.81	1.35
Total capital (billion RMB)	2,440.19	8,955.74	0.27
of which: average net value of fixed assets	951.57	4,413.69	0.22
Profits per 100 RMB capital (RMB)	8.10	1.60	5.06
Profits and taxes per 100 RMB capital (RMB)	12.90	6.20	2.08
Sales Revenue (billion RMB)	3,828.40	6,414.89	0.60
Profits per 100 RMB sales revenue (RMB)	5.10	2.30	2.22
Profits and taxes per 100 RMB sales revenue (RMB)	8.20	8.60	0.95
Sales revenue per 100 RMB capital (RMB)	156.90	71.60	2.19

Source: Chen et al. (1999).

Note: RMB: Renminbi, the Chinese currency, equivalent to $0.125.
For this study, SOEs (State-Owned Enterprises) include all NonTownship and Village Enterprises, but is comprised almost entirely of SOEs.

Similarly, several analysts (e.g., Jefferson and Singh 1999; Lardy 1999) show that TVEs have enjoyed this superior economic performance for at least the past ten years. In their study of factor productivity between SOEs and TVEs, Jefferson, Singh, Junling, and Shouqing (1999:137) reach similar conclusions. Examining total factor productivity performance of China's SOE and TVE sectors between 1980 and 1992, they reach the following conclusions. First, productivity in China's SOE sector increased by about 2 to 4 percent during the period of 1980 to 1992. Second, total factor productivity growth within the TVE sector was about twice that of the SOE sector, and finally, productivity growth differed widely across industries. Productivity growth was typically lowest in extractive industries and highest in light industries, particularly in the electronics industry.

It is apparent from the above discussion then that there is ample evidence to extend the paradox I observed in respect to cokemaking SOE and TVE plants in Shanxi Province to SOE and TVE sectors as a whole for China. In addition, I have given glimpses of what could be the explanation of this paradox, that is, discrepancies in input levels in labor and other materials between the SOE and TVE sectors.

There are a number of economic theories that may explain the above paradox, among them the theory of the firm. In this study, I focus on the theory of ambiguous property rights (Li 1996; Polenske 1999; Weitzman and Xu 1994), because of the findings from our cokemaking surveys that TVEs have very diverse and complex property relationships that do not fit well with standard theories of property. I maintain that informal relationships among the owners, local officials, workers, and administrators allow TVEs to have access to cheaper inputs, and, as such, enjoy better factor productivity, despite the fact that they are less energy efficient than SOEs.

5.4 Theories and Explanations

In this case, I focus on the ownership and governance structure of the TVE sector to explain some of the above paradox. My main argument is that ambiguous property rights, which allow for informal relationships among providers of inputs, such as labor and energy, permit TVEs to acquire them at lower prices, which can then translate into high plant productivity and lower energy-efficiency levels.

Porter, a well-known business economist, points out that there are four factors that define the context for a firm's growth, innovation, and productivity: (1) factor (input) conditions; (2) the context for strategy and rivalry; (3) demand conditions; and (4) related and supported industries (Porter 1998). Here, we focus on the second of the four factors to explain some of the above paradox. Porter's concept "context for strategy and rivalry" refers to the notion that certain firms and sectors have advantages in productivity competition if the context of rules, social norms, and incentives foster sustained investment in forms appropriate to a particular industry or sector (Porter 1998).

TVEs in China are owned by local citizens and controlled by Township and Village Government (TVG), and the TVEs' residual benefits are shared among citizens and TVG officials (Chang and Wang 1994). In general, TVE's communal ownership structures are very complicated, and as a result, have poorly defined property rights. From our own survey of the TVE cokemaking sector in Shanxi Province in 1998, we found that of the 158 cokemaking plants covered in the survey, the managers reported that 37 percent are self-owned; 27 percent are shareholders; 17 percent are village owned; 10 percent are town-owned; 10 percent are joint-owned, including those with foreign firms; 3 percent are rented or leased, and 3 percent are other (Polenske, Chen, Pan, Yang, and Shirvani-Mahdavi 1999). Given the above complexities of ownership structure among TVEs, we might ask, how do TVEs, defined as vaguely defined cooperatives, seem to perform so well?

The essential arguments behind the logic of ambiguous property rights and their relationships to superior economic performance comes from Li (1996), Polenske (1999), and Weitzman and Xu (1994). Li argues that the immature market environment in China, which he calls the gray market, makes ambiguous property rights more efficient than unambiguously defined private property rights (1996). In this view, a gray market is one in which transactions may be blocked due to remnant government regulations. However, a government bureaucrat or a government agency can properly work around the obstacles and make the transactions possible. Thus, the gray market gets its name due to uncertainty regarding whether the transactions will be white (normal market) or black (underground market) (Li 1996). Li argues that when facing a gray market, the entrepreneur may want to include the government as an ambiguous owner. In other words, the arrangement of ambiguous property rights is a response to the grayness of the market, which is a form of market imperfection (Li 1996). This is apparent in the intimate relationships between TVEs and TVGs.

Polenske (1999) says that ambiguous property rights may have helped the cokemaking TVEs to get better access to inputs and at lower prices than SOE cokemaking enterprises. In the case of the cokemaking TVEs in Shanxi Province, most of them are in the countryside and close to coal mines, one of the major inputs as well as labor (peasants).

Concerning the TVE sector in China, policy makers report that transactions are often based on oral agreements instead of written contracts. Even in the case of written contracts, it is often the case that the contracts are incomplete and unspecified in items, or there is no specific punishment for breaching the contract. This is particularly important because part of the popularity of this kind of practice is the importance of long-term relationships and connections for TVE transactions (Liu 1989). Given the importance of long-term relationships and connections, when there are disputes, many TVEs would rather settle privately instead of relying on the courts because they care more about keeping long-term connections, even though doing so may hurt their business in the short run (Cai 1990).

Finally, Polenske (1999) expands on the ambiguous property rights argument by showing that, in addition to the above factor, the reasons for the success of the TVE sector are from a "combination of these control rights with the particular governance mechanisms and economic, social, and political power structures that exist" (Polenske 1999). As such, she argues that three major changes in institutions have had an important effect on TVEs in China. First, the property-right systems that are developed are very complex, rather than a simple transformation from state ownership to private ownership. Second, property rights under each of the diverse ownership structures are usually ambiguously defined. Third, the gift economy, defined as personal exchanges and circulation of gifts, favors and banquets, is enabling local officials to affect both the production and consumption patterns of goods and services (Polenske 1999).

However, it is also important to note that some analysts (e.g., Shleifer and Vishny 1997) have dismissed the notion that ambiguous property rights have had a positive effect on TVEs' profitability. Some analysts claim that although TVEs' property rights may be ambiguous, they are nonetheless better defined than those of SOEs. From their perspective, the fact that TVEs enjoy greater profitability and growth has more to do with the public nature of ownership by the SOEs than the fact that ambiguous property relationships among the TVEs can be responsible for their economic performance. The emphasis is on the fact that property relationships affect the governance structure of the firm. The governance structure of a firm refers to "the ways in which suppliers of finance to the firm assure themselves of getting a return on their investment" (Shleifer and Vishny 1997, p. 751). This is not a very concrete definition of governance structure.

In addition to finance, issues of who has decision rights exist. The governance structure of the SOEs is less defined than TVEs, because the State, in the role of financing SOEs, has no assurance to get adequate returns on its investments, while the governance structure of TVEs is better defined and appears to be much more effective. That is because the main suppliers of investment to the TVEs are the township and village households and outside creditors, who have all the incentives to make sure that their investments will not be appropriated (Steinfeld 1998; Jefferson 1998; Perotti, Sun, and Xu 1999).

The second group looks to Coase (1960) to show that the existence of a property rights market is critical to ensure enterprise efficiency. Coase argued that the means to remedy the inefficient use of public goods is to assign property rights clearly and eliminate transaction costs so that assets can be traded among the individuals or groups who can efficiently use them (Coase 1960). In effect, Coase argued for solving the public-goods problem by creating a property-rights market. From this point of view, absent an effective central contracting agent, the firm assumes the two properties of a public good: non-excludability and non-diminishability. In the absence of an outright change in ownership, to what extent has managerial reform created a structure in which a central contracting agent has the authority and incentive to monitor the firm effectively. As such, according to Coase, ambiguous property relationships among the TVEs are not responsible for their economic performance, and are, in fact, hindering their ability to perform even more efficiently than at present.

When comparing input prices between SOEs and TVEs in China, we note that the average labor wages for TVEs are significantly lower than SOEs. Nationally, the average wage for TVE employees is 4,512 RMB per year, while for SOEs, it is 6,747 RMB per year (1995). This holds true for all industrial sectors (CSY 1999). Second, welfare costs: on average SOEs spend 490 RMB per worker per year in welfare costs, compared to 149 RMB for TVEs (CSY 1999). Similarly, pension costs are higher in the SOE sector than in TVE sector (Jefferson 1999). In the end, it is beyond our scope in this chapter to make a clear empirical connection between ambiguous property relationships and the performance of TVEs, but we believe that the above-observed property relationships have allowed TVEs to enjoy lower input prices than SOEs.

5.5 Conclusion

Virtually all productivity studies of SOEs and TVEs in China during the 1980s and 1990s conclude that productivity in the TVE sector has grown faster than in the SOE sector. These analysts have not, however, attempted to compare the paradoxical nature of TVEs superior economic performance as compared to their inefficient utilization of energy. In this chapter, I show that a majority of TVEs sectors exhibit the paradoxical characteristics of being less energy efficient and more profitable than SOEs in both Shanxi Province and China as a whole. Using Structural Decomposition Analysis, I show that 28 out of 29 TVE sectors in Shanxi Province and China as a whole are less energy efficient than their SOE counterparts. This occurs despite the fact that TVEs in China have enjoyed a far better economic performance than SOEs. In order to reconcile this paradox, I examine the direct and indirect labor and other material inputs between the TVE and SOE sectors in China and Shanxi Province. In order to explain the differences in direct and indirect energy, labor, and materials inputs between SOEs and TVEs, I have extended previous theoretical discussions of ambiguous property rights in the TVE sector to explain some of the findings in this study. My primary argument is that vaguely defined relationships among owners, workers, and local administrators allow TVEs to establish informal contract arrangements that give them access to cheaper inputs.

References

Bell, Michael W., Hoe Ee Khor, and Kalpana Kochlar. 1993. China at the Threshold of a Market Economy. *International Monetary Fund (IMF) Occasional Paper*, No. 107, Washington DC: IMF.

Cai, Jinyun. 1990. A Study on Legislation on TVEs. In *A Collection of Studies on Reform*. Beijing: Zhongguo Zhuoyue Publishing Corporation: 195-207.

Chang, Chun and Wang Yijiang. 1994. The Nature of the Township-Village Enterprise. *Journal of Comparative Economics* **19**(3): 434-452. March.

Coase, Ronald H. 1960. "The Problem of Social Cost." *Journal of Law and Economics*, 3, 1-44.

Jefferson, Gary H. 1998. China's State-Owned Enterprises, Public Goods and Coase. *American Economic Review* **88**(2):428-432. February.

Jefferson, Gary H., and Inderjit Singh (eds.) 1999. *Enterprise Reform in China: Ownership,Transition, and Performance*. New York: Oxford University Press.

Jefferson, Gary H., Inderjit Singh, Xing Junling, and Zhang Shouqing. 1999. China's Industrial Performance: A Review of Recent Findings. In *Enterprise Reform in China: Ownership, Transition, and Performance.* Gary H. Jefferson and Inderjit Singh, Eds. New York: Oxford University Press: 127-152.

Lardy, Nicholas R. 1998. *China's Unfinished Economic Revolution.* Washington DC: Brookings Institution Press.

Li, Daokui. 1996. A Theory of Ambiguous Property Rights in Transitional Economies: The Case of Chinese Non-State Sector. In *Economic Study: Collection of Papers*, edited by Chen Zhangwu, Beijing, People's Republic of China: Tsinghua University, China Economies Studies Center, Tsinghua University Press: 130-158.

Liu, D. 1989. "On Current Disputes of TVE Contracts and their Resolutions." *Rural Economy*, 8: 89-101.

Perotti, Enrico C., Laixiang Sun, and Liang Zuo. 1999. State Owned versus Township and Village Enterprises in China. *Comparative Economic Studies*, **41**(2/3): 151-179.

Polenske, Karen R. 2002. "Taking Advantage of a Region's Competitive Assets: An Asset-Based Regional Economic-Development Strategy." In *Entrepreneurship, Firm Growth, and Regional Development in the New Economic Geography.* Trollhätten, Sweden: Universities of Trollhätten,/Uddevalla (Papers from Symposium 2000, June 15-17): 527-544.

Polenske, Karen R., Ali Shirvani-Madavi, Wei Guo, Yang, Cuihong, and Xikang Chen. 1999a. A Field Trip to Shanxi Province, China. *Multiregional Planning Group.* Department of Urban Studies and Planning, Massachusetts Institute of Technology. (January).

Porter, Michael E. 1998. *On Competition.* Boston: Harvard Business School Press.

Shleifer, Andrei, and Robert W. Vishny. 1997. A Survey of Corporate Governance. *Journal of Finance*, **52**(2): 737-783. February.

Song, Linda, and He Du. 1990. The Role of Township Governments in Rural Industrialization. In *China's Rural Enterprises: Structure, Development, and Reform*, edited by William Byrd and Qingsong Lin. Oxford, UK: Oxford University Press: 342-357.

SSB (State Statistical Bureau of China). 1997. *China Statistical Yearbook, 1998* (English Edition). Beijing: China Statistical Publishing House.

SSB (State Statistical Bureau of China). 1999. *China Statistical Yearbook, 1998* (English Edition). Beijing: China Statistical Publishing House.

SSB (State Statistical Bureau of China). 2003. *China Statistical Yearbook, 1998* (English Edition). Beijing: China Statistical Publishing House.

Steinfeld, Edward S. 1998. *Forging Reform in China.* Cambridge: Cambridge University Press.

Sun, Laixiang. 1998. Estimating Investment Functions Based on Cointegration: The Case of China. *Journal of Comparative Economics.* **26**(1): 175-191, March.

Weitzman, Martin L., and Chenggang Xu. 1994. Chinese Township-Village Enterprises as Vaguely Defined Cooperatives. *Journal of Comparative Economics* **18**: 121-145.

Xiao, Geng. 1991. Managerial Autonomy, Fringe Benefits, and Ownership Structure. *Research Paper Series, No. 20*, Socialist Economies Reform Unit, Country Economies Department, World Bank.

Yang, Cuihong, and Xikang Chen. 2000. Study on Industry Structure Change for Sustainable Development of China Township-and-Village Enterprises. *Multiregional Planning Group Working Paper* [Draft] Department of Urban Studies and Planning, Massachusetts Institute of Technology.

Zou, Liang, and Laixiang Sun. 1998. A Theory of Risk Pooling and Voluntary Liquidation Firms: With an Application to Township and Village Enterprises in China. *Tinbergen Institute Discussion Paper*, TI 98-123/2. Amsterdam: Tinbergen Institute.

---

[1] At time of writing, Multiregional Planning (MRP) Research staff, MIT, Cambridge, MA, USA; currently Strategy Consultant, Accenture, Minneapolis, MN.

CHAPTER 6

MODELING COST AND POLLUTION OF COAL AND COKE TRANSPORTATION IN SHANXI PROVINCE

Yan CHEN,[1] Steven KRAINES,[2] and Karen R. POLENSKE[3]

6.0 Introduction

Throughout this book, we show that the cokemaking industry in Shanxi Province is highly polluting and energy-intensive. Planners have traditionally focused only on ways to reduce cokemaking pollution and energy consumption. However, we have found that the particulate emissions and energy consumption of the diesel trucks that transport coke and coal have a similar level of importance. In light of this discovery, we have developed measurements and planning tools that help us explore options for overall cost and pollution minimization in both cokemaking plants and transport of coal and coke. We describe these developments in this chapter.

In the cokemaking sector, the supply chain runs from: (1) coalmines to (2) the transportation of coal to (3) cokemaking plants to (4) the transportation of coke to (5) coke consumers (Figure 6.1). Although the cokemaking plant is the core component of this supply chain, coal and coke transportation play important economic and environmental roles as well. In Shanxi Province, which is an inland mountainous region, transportation costs alone account for approximately one-third of total production costs (MRP 2001). Coal and coke transportation costs are affected by choice of transport routes and modes, the location of the cokemaking plants, and also accessibility to coalmines and coke consumers.

Karen R. Polenske (ed.), The Technology-Energy-Environmental-Health (TEEH) Chain in China: A Case Study of Cokemaking, 91–108.

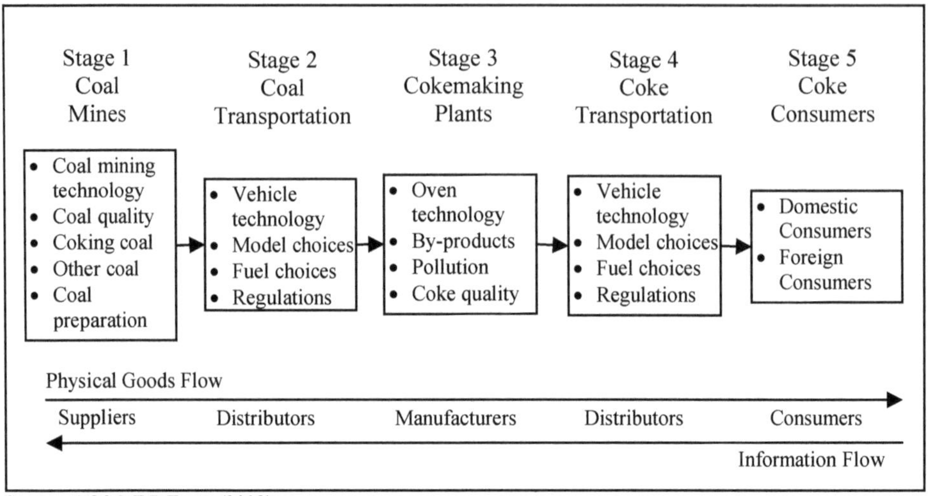

Source: AGS MRP Team (2002).

Figure 6.1: Cokemaking Supply Chain in Shanxi Province

Currently, national environmental regulations are forcing cokemaking plant managers to close inefficient and highly polluting small-capacity cokemaking plants. After the State Economic and Trade Commission, State Environmental Protection Agency, and former Ministry of Machinery issued the #367 directive on June 5, 1997, the Shanxi government began closing small-capacity plants, i.e., those with production capacities smaller than 200,000 tonnes per year. Most of these small-capacity plants used indigenous and modified indigenous coke-oven technologies (Chapter 3). The Shanxi government has decreed that these small indigenous plants should be replaced either with large-machinery cokemaking plants or the so-called "clean" cokemaking plants, whose annual capacity is usually larger than 200,000 tonnes (Chapter 3). In addition, as of 2003, the Shanxi government has indicated that it plans to consolidate its cokemaking industry by building several cokemaking industrial parks in the province. Consequently, plant managers need either to enlarge the capacity of the existing plants that meet the environmental regulations or build new large-capacity, environmentally friendly plants.

These government directives have resulted in both an increase in the average plant capacity and the adaptation of new technologies (Chapter 2). Different coke-oven technologies vary greatly in their respective investment and land costs. For example, clean coke ovens have low operational and investment costs, but land-use requirements are relatively large, resulting in high land costs. Although the use of large-machinery coke ovens usually increases operational and investment costs, less space is required, lowering land costs.

In addition, from the perspective of transportation costs and pollution, significant increases could occur in the distances that coal and coke must be transported when a large number of widely scattered plants are replaced with a small number of large plants.

With this type of plant siting, we describe with the use of a location-economics model how the optimal siting of plants will affect transportation costs. For example, if there is a large coke consumer located between two mines, the economics of the location decision might favor the construction of two cokemaking plants on either side of the consumer. However, when one plant is closed and the other enlarged in order to decrease the marginal production costs, the transportation costs will increase due to the added transport distance. Thus, one key question when considering the replacement of many small cokemaking plants with a few large plants is "how does the increase in transportation cost compare with the decrease in the plant production cost?" This question leads us to the following hypothesis.

6.1 Hypothesis

Our working hypothesis is as follows: replacement of widely scattered small-capacity plants with a small number of large-capacity cokemaking plants in industrial parks will result in a significant increase in the total cost, energy consumption, and pollution emissions along the cokemaking supply-chain in Shanxi Province due to the increased need for transportation. We test this hypothesis using the methods and analytical tools described in Section 6.3.

6.2 Background

We explain first how we adapt geographic information systems (GIS) technologies and draw upon industrial-location theories to create the unique analytical methods we use to test our hypothesis.

6.2.1 GIS-based Planning-Support Systems

With the advances in information technology and database management, many analysts have begun using Geographic Information System (GIS) technology for planning and decision-making. Planners can use GIS as a database management system to perform quantitative analyses based on digital maps. A GIS-based Planning Support System (GPSS) is a computer-based system that uses GIS technologies in an organized way to help decision makers and policy analysts conduct spatial information processing and studies in fields such as transportation and environmental studies (Gittings, Sloan, Healey, Dowers, and Waugh 1993).

6.2.2 Industrial-Location Theories

Although the free-market location theories used in the West are not completely applicable to the still partially planned economy in China, we believe that modifications of industrial-location theories could help analysts investigate the optimality of potential locations for coke plants. DiPasquale and Wheaton (1996) show that at least in Western countries, land rents tend to be high near urban centers and tend to decrease further way from those centers (Figure 6.2). Thus, industrial firms often locate in suburban, or even rural, areas. We find that with modification, we can interpret the location decisions occurring in China using some of the same location theory rationale.

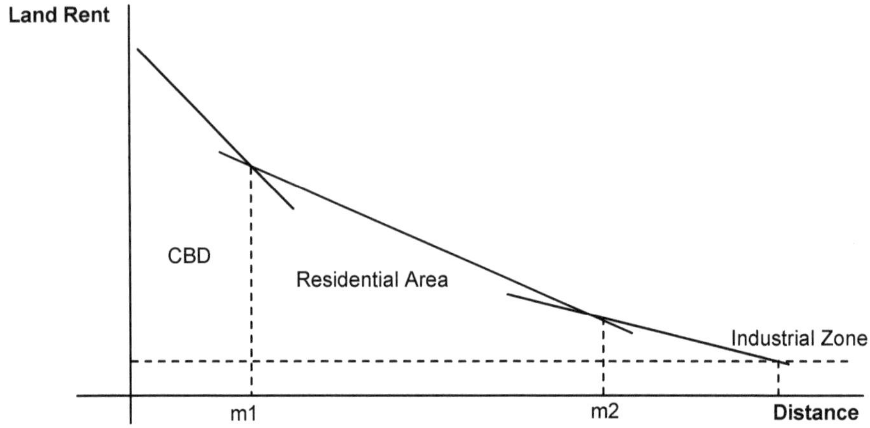

Source: DiPasquale and Wheaton. (1996:102).
Note: CBD—Central Business District

Figure 6.2: Land Rents for Commercial, Residential, and Industrial Uses.

DiPasquale and Wheaton (1996) also indicate that during the twentieth century, changes in both production and storage methods have greatly increased the amount of land used per unit of output by industrial firms in the West. Even in the same industry, industrial firms with different production technologies often have widely different land-use requirements. These different requirements can cause firms with different production technologies to have significantly different land-rent gradients and therefore to choose different locations. Although, as we noted above, neoclassical location theories cannot be applied to Chinese location decisions without some modification, some of the land-use phenomena predicted by those theories are in fact occurring in China. For example, transportation costs are an important factor in determining industrial locations in China. Many cokemaking plants in Shanxi Province are started by the peasants who live in rural areas and can take advantage of the proximity of local coal mines, often within 25-75 kilometers from the peasants' village, which results in low transportation costs.

The #367 directive is creating a fundamental tension between economies of scale and the impact of distance for rural coke producers. Economies of scale normally would favor a few large, widely separated plants, which has not occurred in Shanxi Province. As we explained earlier, small plants located in a dispersed spatial network generally have lower distance-related (transportation) costs than the larger plants. In other words, according to neoclassical economic theory, economies of scale are expected to make larger plants more economically efficient than small ones, and, in aggregate, easier to manage. However, in the actual situation of rural coke production in Shanxi Province, the higher transportation costs, larger single investments, capital constraints, reduced flexibility, and the locational disadvantages of large plants appear to have favored the establishment of smaller plants. The key policy issue therefore is to balance these opportunity costs correctly (Rees and Stafford 1986).

The question of what size and number of cokemaking plants in Shanxi Province results in the lowest total cost forms the basis for the research work we describe in the following sections. With the use of the GIS technologies and a network optimization algorithm, we have conducted transportation and industrial-location analyses to determine the equilibrium point between dispersed and aggregated cokemaking firms for our case study of the coal and coke transportation in Shanxi Province.

6.3 Methods and Project Design

We designed the Shanxi Province Geographic Planning Support System (SPGPSS) and applied it to the analysis of transportation and location choices faced by the coke managers and local government officials in Shanxi Province, China, in two stages.

During the first stage of the development of the SPGPSS (1999-2000), Steven Kraines and Takeyoshi Akatsuka created a systems tradeoff model that linked a GIS transportation network database for Shanxi Province with the locations and capacities of coal mines, cokemaking plants, and coke consumers, including coke exported from the province (Akatsuka 2001; Kraines, Akatsuka, Crissman, Polenske, and Komiyama 2003). Using components developed for that systems tradeoff model, Chen Yan constructed an extended GPSS system that incorporated updated location and capacity data (Chen 2003). Although we refer to the SPGPSS as our model, various members of the AGS cokemaking team as well as other researchers working with us have made important contributions since 1997 towards the development of the SPGPSS that we describe.

Using the SPGPSS, we compare three location scenarios in terms of the minimized total costs, energy consumption, and emissions from transportation and cokemaking plants. Then, we explore the underlying reasons behind those differences; evaluate several coke-oven technology options for the plants in terms of plant costs and emissions; measure how the new highway construction affects the coal and coke transportation by reducing costs, energy consumption, and emissions; and, finally, recommend the optimized plant location, capacity allocation, transportation routes and modes by comparing the results from these alternatives.

The SPGPSS has four major components: (1) database, (2) map viewer, (3) scripts, and (4) numerical models. Each component has a special function and is independent of other components, but all of the components are closely connected. We show the basic structure of this SPGPSS in Figure 6.3.

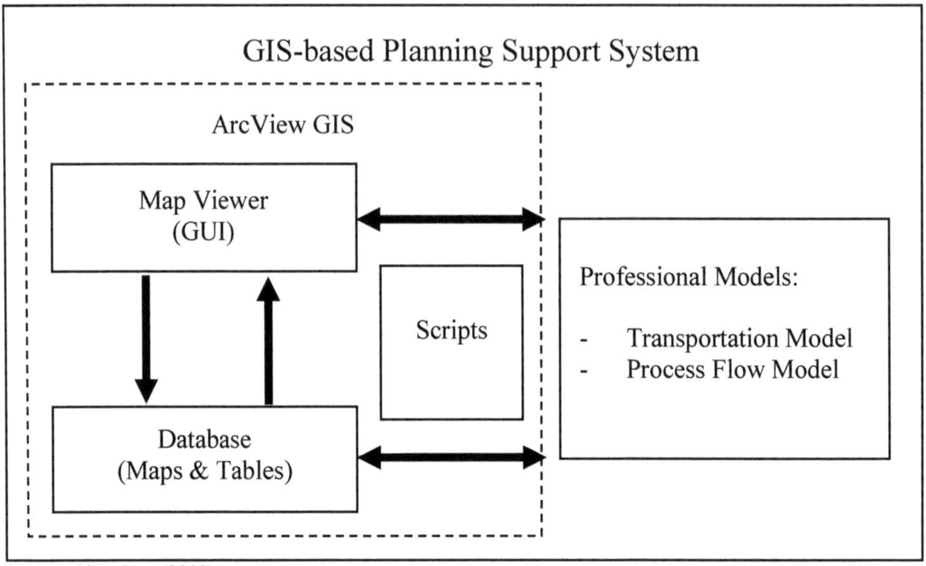

Source: Chen Yan (2003).

Figure 6.3: Structure of GIS-based Planning Support System (GPSS)

6.3.1 Database–Core of SPGPSS

We formed the core of the SPGPSS with a database of the locations of cities, towns, villages, mines, major coke consumers such as steelmaking plants, and the cokemaking plants that are currently in operation. In addition to locational data, the database contains attribute data, such as city population and plant-production rates, network data of the road and railway transportation infrastructure, and boundary data delineating cities and counties.

6.3.2 Map Viewer–Graphic User Interface (GUI) of SPGPSS

We used the ArcView GIS software package to display the results of our analysis based on the model and spatial data (ESRI 1997). Figure 6.4 shows a screen capture of the ArcView software interface displaying the coalmines, cokemaking plants, railways, roads categorized by the service levels, cities and towns, and the border of Shanxi Province from the viewpoint of coal transportation, i.e., transportation from coalmines to cokemaking plants.

6.3.3 Scripts and Models

Scripts are the bridges that connect models with the GIS software, thereby integrating the different components into one system. Using our own scripts, we have connected the SPGPSS system with two numerical models: the transportation NETFLOW model (Kraines et al. 2003) and the process-flow model (Polenske and McMichael 2002). The SPGPSS is organized as shown in Figure 6.5.

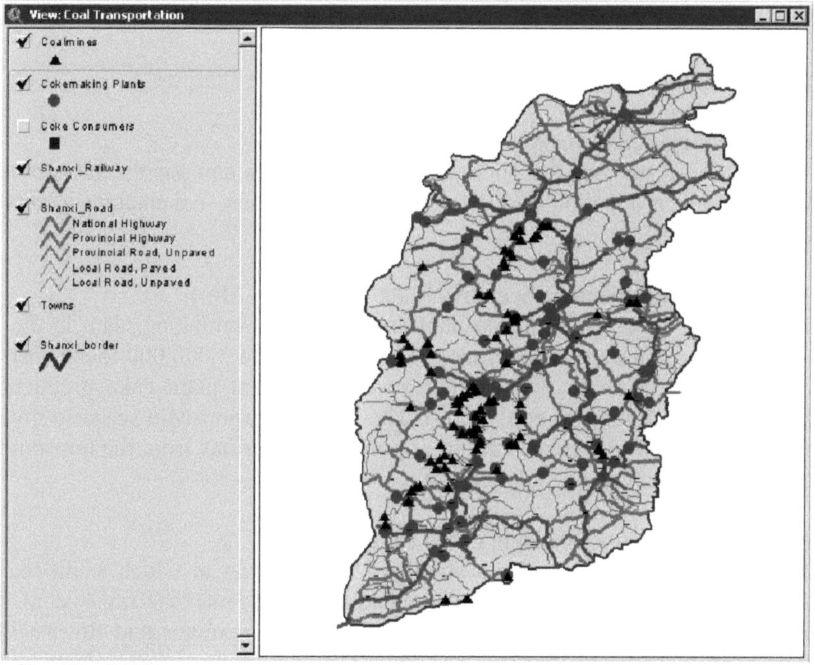

Source: Chen Yan (2003).

Figure 6.4: View of Coal Transportation in Shanxi Province, China, 2000

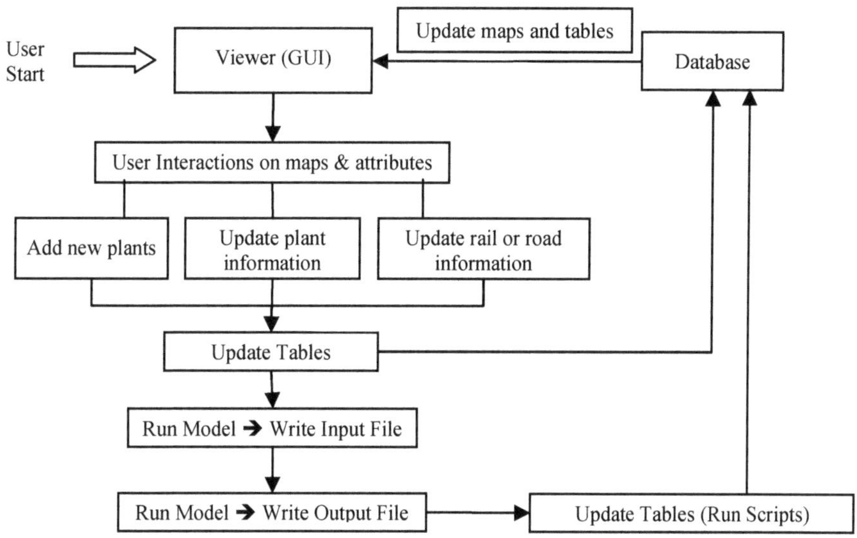

Source: Chen Yan (2003).

Figure 6.5: System-Flow Chart of SPGPSS

6.3.4 Three Location Scenarios

We consider the following three location scenarios for these analyses:

2000 Base Scenario (2000 Base)

This scenario comprises the optimized transport routes and supply flows based on the year 2000 production and distribution of coalmines, cokemaking plants, and coke consumers.

2000 Transport-Minimization Scenario (2000 Transport-Min)

This scenario is based on the assumption that each cokemaking plant in the 2000 Base Scenario can expand its production capacity up to 3,000,000 tonnes per year, which we assume to be the upper limit for individual plant coke production in Shanxi Province (Kraines et al. 2003). The 2000 Transport-Min scenario gives the minimum cost, energy consumption, or pollution emissions from the transportation stages in the cokemaking supply chain.

2000 Plant-Minimization Scenario (2000 Plant-Min)

The plant-minimization scenario simulates the situation in which small-scattered plants are closed and replaced by aggregated large-scale plants (Kraines et al. 2003). With this scenario, we provide the optimized transport routes and flows with the assumption that Shanxi Province will have just enough large-scale plants, all with 3,000,000 tonnes per year capacity, to satisfy the total coke demand. This scenario gives the lowest cost, energy consumption, or pollution emissions for the cokemaking plants that can be achieved through economies of scale.

6.4 Analyses

Based on the year 2000 production and distribution data of coalmines, cokemaking plants, and coke consumers in Shanxi Province (China Statistics Press 2001), we conducted analyses of alternatives to minimize:

1. Total costs from transportation and cokemaking plants;
2. Total energy consumption from transportation and cokemaking plants;
3. Total pollution emissions from transportation and cokemaking plants.

For each minimization, we used SPGPSS to test the three scenarios described in the previous section: 2000 Base Scenario, 2000 Transport-Minimization (Transportation-Min) Scenario, and 2000 Plant-Minimization (Plant-Min) Scenario.

6.4.1 Minimization of Total Cost

We calculated the transportation cost for each scenario by using SPGPSS to find the optimal transportation routes and flows for the three scenarios. We used a plant-cost model to estimate the plant cost based on plant-production data in 2000 (Chen 2000). The total cost is the sum of transportation and plant costs.

Tables 6.1 and 6.2 and Figures 6.6 and 6.7 show the results for the scenario options where all the plants use the "clean" non-recovery coke-oven technology and where all the plants use large-machinery coke-oven technology, respectively, as described in Chapter 3.The results show that the 2000 Plant-Min Scenario, the scenario that is characterized by a small number of large-capacity plants in Shanxi Province, has the lowest total cost in both cases of coke-oven technology. Although the transportation costs in the Plant-Min Scenario are the highest among the three scenarios, the considerably lower plant costs in this scenario more than compensate for the increase in total costs from transportation. In the case of clean coke-oven technology, the plant cost in the Plant-Min Scenario is only 50% of the plant cost in the Transport-Min Scenario, the scenario where plants are sited and sized in order to minimize transportation costs, and 89% of the plant cost in the Base Scenario, the scenario that is intended to show the current situation in the province.

The model calculation results in Tables 6.1 and 6.2 show that for all of the scenarios, plant costs comprise more than 80% of the total cost for the clean coke-oven technology and more than 95% of the total cost for the large-machinery coke-oven technology. Because the contribution of transportation costs is almost negligible, the scenario with the lowest plant cost, i.e., the Plant-Min Scenario, has the lowest total cost. However, it is possible that this is a consequence of the assumptions that we have used in our cost models. A small change in the plant-costing model parameters that control the economies of scale for the cokemaking ovens could cause this result to be reversed. Therefore, policy makers need to evaluate the economies of scale for the different cokemaking oven technologies carefully when applying these results to actual decision making.

Table 6.1: Total Cost of Clean Coke-Oven Technology

Unit: Million Renminbi

Scenarios	2000 Base		2000 Transport-Min		2000 Plant-Min	
	Cost	Percentage of total cost	Cost	Percentage of total cost	Cost	Percentage of total cost
Transportation cost	1,104	15%	574	5%	1,341	19%
Plant cost	6,456	85%	11,536	95%	5,768	81%
Total cost	7,560	100%	12,110	100%	7,109	100%

Source: Chen Yan (2003).

Table 6.2: Total Cost of Large-Machinery Coke-Oven Technology

Unit: Million Renminbi

Scenarios	2000 Base		2000 Transport-Min		2000 Plant-Min	
	Cost	Percentage of total cost	Cost	Percentage of total cost	Cost	Percentage of total cost
Transportation cost	1,104	3%	574	1%	1,341	5%
Plant cost	33,251	97%	94,403	99%	26,351	95%
Total cost	34,355	100%	94,977	100%	27,692	100%

Source: Chen Yan (2003).

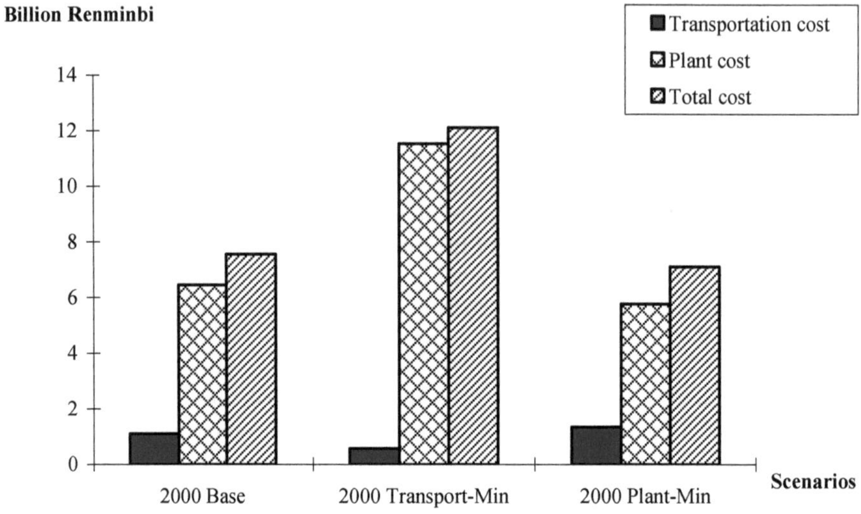

Source: Chen Yan (2003).

Figure 6.6: Total Cost of Clean Coke-Oven Technology

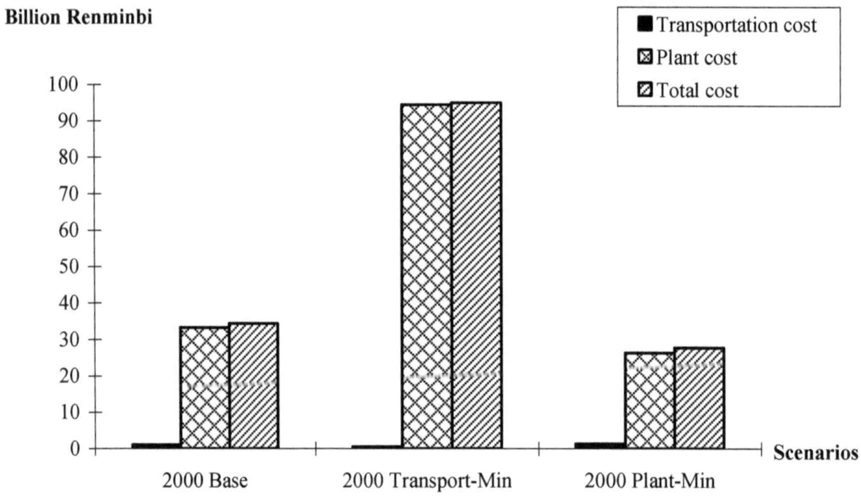

Source: Chen Yan (2003).

Figure 6.7: Total Cost of Large-Machinery Coke-Oven Technology

In summary, from the perspective of total cost minimization, the model calculations indicate that the Plant-Min Scenario is the best. This result negates the hypothesis we stated earlier that replacing the distributed small-capacity plants with large-capacity cokemaking plants would increase the total cost from transportation and the cokemaking process.

Next, we analyze the results of the total cost minimization in detail.

6.4.1.1 Transportation Flow and Cost Comparisons

First, we compare the optimization model results of road transportation with rail transportation. In the 2000 Base Scenario, in terms of transport flows, road transportation accounts for 75% of the total transport flows, and rail transportation accounts for 25%. The percentage of total transportation cost that is accounted for by road transport cost is 78%, and railway transport cost is 22% of the total cost. The difference between the percentages for transport flows and transport costs is due to the higher cost of transportation by roads per unit of coal or coke transported. While rail transportation comprises 25% of the total transport flows, it only accounts for 22% of the total costs, which implies that railway transportation is, on average, slightly less expensive than road transportation in this scenario. The larger fraction of coal and coke transported by road is due to the limited capacity of the railway transportation that is assumed in the model (Akatsuka 2001).

Next, we compare the transportation flows and costs for coal and coke. In the 2000 Base Scenario, the coal transport flow makes up only 20% of the total transport flows. The optimized arrangement generated by the SPGPSS tends to locate cokemaking plants nearer to the coalmines than to the coke consumers, so that the total flows of coal are less than the total flows of coke. The coal transport cost is 27% of the total transport cost, so the average cost per unit flow is slightly higher than the cost for coke. The higher average cost may be a reflection of the greater reliance of coal transportation on the higher cost road transport, indicating that the transport flows and costs may be effectively reduced even more if the cokemaking plants can be relocated even nearer to the coalmines.

6.4.1.2 Plant-Cost Comparison

The determining factor of plant cost for the clean coke-oven technology is the operating cost, which accounts for 96% of the total plant cost. The land cost contributes insignificantly (0.1%) to the total plant cost, perhaps partially because of the rural location of many of the coke plants.

6.4.2 Minimization of Total Pollution Emissions and Energy Consumption

To estimate the minimization of the pollution emissions and energy consumption, we ran the SPGPSS under the three scenarios and calculated the transportation pollution emissions and energy consumption for each scenario. During field trips to Shanxi Province (MRP 1998, 1999, 2000, 2002, 2004), members of our research team found that the PM_{10} emissions from heavy-diesel trucks are even greater than the PM_{10} emissions from the cokemaking plants except directly at the quenching car. Thus, PM_{10}

emissions are appropriate for comparing the pollution contributions from transportation and cokemaking plants. We used the PM_{10} and sulfur dioxide (SO_2) as the major measurements for emissions both from transportation and cokemaking plants. We have not included nitrogen oxides in our analysis due to lack of data for plant emissions.

Compared with the other two scenarios, the plants in the Transport-Min Scenario are closer to the suppliers, consumers, and major transportation routes, which results in the minimum transportation distance. Therefore, the Transport-Min Scenario gives the smallest transportation related PM10 and SO2 emissions among the three scenarios, as well as the lowest transportation energy consumption. Furthermore, because our data for pollution emissions, even SO2 emissions, are much higher for the transportation stage than the cokemaking stage, the Transport-Min Scenario also gives the smallest total pollution emissions and energy consumption.

Unlike the cost calculations presented earlier, these calculations for pollution and energy support our hypothesis that the replacement of the scattered small-capacity plants with large plants will increase the total emissions and energy consumption from transportation and cokemaking plants. However, again we stress that these results are highly sensitive to the specification of the different models for pollution emissions and energy use at the plant production and transportation stages of the cokemaking supply chain. For example, we have assumed that there are no economies of scale for pollution and energy in the plants. If pollution and/or energy use in plants have significant economies of scale, the replacement of small-capacity plants with large plants will tend to have lower total emissions and energy consumption than we have shown here.

6.4.3 Coke-Oven-Technology Impact Analysis

The type of coke-oven technology that a plant manager selects can have large impacts on the plant cost, energy consumption, and pollution emissions. We have studied three scenario options for coke-oven technology:

Option 1: All plants in the province use clean coke-oven technology.
Option 2: All plants in the province use large-machinery coke-oven technology.
Option 3: Plants whose capacity is more than or equal to 500,000 tonnes use large-machinery coke-oven technology; plants whose capacity is less than 500,000 tonnes use clean coke-oven technology.

We assume that the total annual production of coke from all of the cokemaking plants combined is the same in each of these three options. We found that the plant cost in Option 1 is the lowest of the three options and is about one-fifth of the plant cost in Option 2 (Table 6.3). The major reason for this large difference is the lower operation and investment costs of the clean technology. Although the land cost of clean plants is higher than that of the large-machinery plants, as we noted earlier, land accounts for less than 0.1% of total cost for either technology Table 6.3.

Of the different components of plant cost, the operation cost is the largest cost for all of the options (Figure 6.8). As shown in Table 6.3, land costs are extremely small, so they do not show in the figure. Note that Option 1, the clean coke-oven technology, features the highest percentage of operational cost in the total plant cost, mainly because it is less capital intensive than the large-machinery coke ovens.

The investment cost is 4% of the total plant cost in Option 1. This percentage increases to 26% in Option 2. In Option 1, because all the by-products (tar, ammonia, etc.) are fully combusted in the clean coke ovens, plants do not need to purchase and install the equipment and facilities for by-product recovery and pollution abatement. By contrast, in Option 2, the large-machinery technology requires initial investment inputs for the advanced machinery equipment and facilities for by-product recovery and pollution abatement.

Table 6.3: Plant-Cost Comparison of Three Coke-Oven Technology Options

Unit: million Renminbi/year

	Operation Cost	Investment Cost	Land Cost	Plant Cost
Option 1 (Clean technology)	6,206	244	6	6,456
Percentage of plant cost	96.2%	3.8%	0.1%	100.0%
Option2 (Large-machinery technology)	24,673	8,573	4	33,250
Percentage of plant cost	74.2%	25.8%	0.0%	100.0%
Option 3 (Clean and large-machinery)	18,512	4,730	5	23,247
Percentage of plant cost	79.6%	20.4%	0.0%	100.0%

Source: Chen Yan (2003).

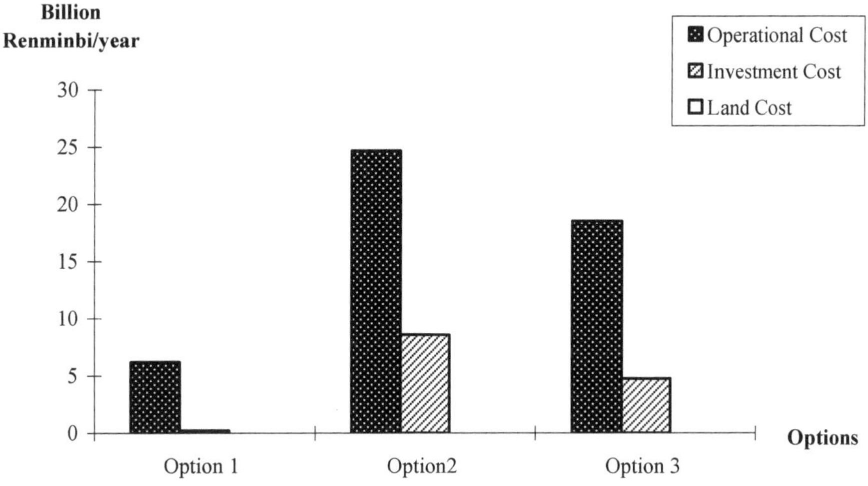

Source: Chen Yan (2003).

Figure 6.8: Plant-Cost Comparison of Three Coke-Oven Technology Options

6.4.4 New-Highway-Construction and Speed-Improvement Impact Analyses

The highway system is developing quickly in Shanxi Province. We obtained the information presented here concerning future highway plans from an interview with an official in the Shanxi Province Transportation Planning Department. (AGS MRP Field Trip Interviews 2002). The dates and the exact origin and destination of the highway, of course, are subject to change. With Taiyuan as the transportation hub, the highways in Shanxi Province form a road network linking all the counties in the province. In the North-South direction, the Datong-Yuncheng Highway is a major highway connecting the cities in the south and north of the province. The new Taiyuan-Changzhi-Jincheng Highway will be built from 2005 to 2008. In the West-East direction, the Taiyuan-Jiuguan Expressway, which joins the Beijing-Shijiazhuang expressway, connects Beijing-Tianjin-Tanggu expressway and Beijing-Shenzhen expressway, and leads to Beijing and the region of Bohai Sea directly. This highway will also extend westwards from Taiyuan to the Lishi-Liulin area when several new highways are built in the next five years.

In the Tenth Five-Year Plan, Shanxi Province intends to improve its highway transportation system by investing at least Renminibi (RMB) 10 billion on new highway construction over the next five years (China State Development and Planning Committee 2001). Also, due to improving road conditions, the average travel speed on the roads in towns and villages increased from 30 kilometers (km) per hour in 1990 to 40 km per hour in 2003.

Based on the map of Shanxi Province major highway construction in the tenth five-year plan provided by Shanxi Province Development and Planning Committee, we added two major new highways in the SPGPSS: Taiyuan-Changzhi-Jincheng Highway and Taiyuan-Lishi-Liulin Highway. Furthermore, we increased the speeds of the town-village level roads from 30 km per hour to 40 km per hour in the transportation model. This speed improvement can increase fuel efficiency from 0.056 liter/km-tonne to 0.043 liter/km-tonne for transportation on those roads (Akatsuka 2001). We implemented the SPGPSS model to obtain results for the different scenarios (Tables 6.4 and 6.5).

New highway construction and road-speed improvements result in the decrease of transportation cost, energy consumption, and PM_{10} emissions. Construction of new highways increases the routing options available for the transportation of coal and coke, which tend to decrease both costs and pollution directly by lessening the total kilometers that are traveled. Although we have modeled cost as a function both of travel distance and travel time, we find less clear effects from increases in road speed. We use pollution-emission coefficients that are minimum at travel speeds that are greater than most of the travel speeds we have assumed for the relatively congested roads in Shanxi Province; therefore, in general, higher speeds will lower both costs and pollution. Of the three scenarios, the Transport-Min Scenario is the most effective in taking advantage of new highway construction and road-speed improvements to reduce the cost and emissions, Therefore, the Transport-Min Scenario has the largest decrease and the Plant-Min Scenario has the smallest decrease both for transportation cost and emission pollutions.

Table 6.4: Transportation Cost: Comparisons Before and After the New Highway Construction and Road-Speed Improvements

Unit: Million Renminbi/year

Scenarios	Before	After	Decrease Percentage
2000 Base	1,104	1,082	2.0%
2000 Transport-Min	574	556	3.1%
2000 Plant-Min	1,341	1,322	1.4%

Source: Chen Yan (2003).

Table 6.5: Transportation PM Emissions: Comparisons Before and After the New Highway Construction and Road-Speed Improvements

Unit: Kilogram/year

Scenarios	Before	After	Decrease Percentage
2000 Base	1,622,440	1,596,488	1.6%
2000 Transport-Min	802,219	788,258	1.7%
2000 Plant-Min	2,181,582	2,154,016	1.3%

Source: Chen Yam (2003).

6.4.5 Industrial-Park Location-Choice Analysis

In 2004, Shanxi Government officials designated the following eight "coking industrial garden" regions: Hongtong, Jiexiu, Xiaoyi, Fenyang, Qingxu, Gujiao, Lucheng, and Hejin (Fan 2004: 146). To simplify the analysis, we used the initial plan the Shanxi government officials set forth in 2003 to build two cokemaking industrial parks in the province to increase economies of scale and reduce environmental pollution. The principle approach would remain the same if we were to use eight instead of two parks.

For the two parks, they tentatively selected locations in or near Lishi-Liulin, Linfen, and Jiexiu, which are major cokemaking industrial areas in Shanxi Province. By using the SPGPSS, we made simulations of several location arrangements of those industrial parks and compared the resulting total transportation cost, energy consumption, and emissions calculated by the SPGPSS.

We consider three combinations of cokemaking industrial park locations: (1) Lishi-Liulin and Linfen, (2) Linfen and Jiexiu, and (3) Lishi-Liulin and Jiexiu. For each of these combinations, we have assumed that each industrial park produces half of the current total coke production of Shanxi Province.

Of the three combinations of cokemaking industrial parks, the Linfen-Jiexiu combination has the lowest operating and transportation cost, energy consumption, and PM_{10} and SO_2 emissions from coal and coke transportation mainly because Linfen and Jiexiu are closer to the big coal suppliers and major highways and railways than Lishi-Liulin (Table 6.6). Although the planners undoubtedly will take additional factors into consideration for the final industrial-park location decision, our study illustrates the type

Table 6.6: Comparison of Three Cokemaking Industrial Park Scenarios

Scenarios	Lishi-Liulin & Linfen	Lishi-Liulin & Jiexiu	Linfen & Jiexiu
Cost (RMB/year)	1,885,228,826	2,160,621,076	1,883,268,865
Energy (1000kcal/year)	1,921,549,427	2,079,814,965	1,773,941,756
PM (kg/year)	3,492,874	3,995,512	3,372,174
Sox (kg/year)	2,252,170	2,552,780	2,208,906

Source: Kraines, Akatsuka, Crissman, Polenske, and Komiyama 2003.

of quantitative comparisons that policy makers and plant managers can conduct using the SPGPSS.

6.5 Conclusion

We created a Shanxi Province GIS-based Planning Support System (SPGPSS) for conducting transportation and industrial plant location studies of the cokemaking sector in Shanxi Province. By integrating database, map viewer, scripts, and numerical models in a Geographic Information System (GIS) environment, the SPGPSS can help a regional planner or analyst to optimize plant locations, transport routes, and transport modes under the different scenarios at the provincial level as well as to compute the corresponding cost, energy consumption, and pollution emissions in the transportation process.

Using SPGPSS, we tested the hypothesis that combining plants into several large-capacity plants or industrial parks would significantly increase transportation costs and consequently total costs, energy consumption, and pollution emissions as compared with the current situation where the plants are distributed throughout the region. We discovered that, in fact, from the perspective of total cost minimization, the merged large-capacity cokemaking plants and industrial parks would reduce, rather than increase, the total cost from the transportation and cokemaking process, mainly because average operation costs per tonne would be reduced. From the perspective of energy-consumption and pollution-emission minimization, however, the calculation results indicate that the replacement of dispersed small-capacity plants with large-capacity plants and industrial parks would increase total energy consumption and pollution emissions. We stress that these conclusions are sensitive to the assumptions and model optimizations used in this study. However, we also add that the flexible interfaces provided by the SPGPSS make it easy to test the effects of different assumptions and parameter values on the optimized results.

In addition, we found that, as expected, the type of coke-oven technology used has a great impact on the plant cost, energy consumption, and pollution emissions, which, consequently, can affect the plant location and transportation choices that a plant manager makes. The transportation cost and pollution emissions both decreased after new highway construction and road-speed improvements were completed, especially in the transport-minimization scenario. In the industrial-park location analysis, the simulation results indicate that the choice of the Linfen and Jiexiu cokemaking parks would reduce transportation cost, energy consumption, and pollution more than the other two alternatives due to the Linfen-Jiexiu greater proximity to large coal suppliers as well as major highways and railways.

In order to plan the future of the cokemaking industry in Shanxi Province, coke managers and policy makers need to consider the transportation and coke-plant (capacity and technology considerations) stages of the supply chain and incorporate factors from both into their location and operating decisions. In particular, it is critical for them to consider how best to optimize the location and transport routes for large-capacity plants and industrial parks in order to reduce the overall costs and pollutants from the cokemaking and transportation processes.

References

AGS (Alliance for Global Sustainability). 2002. AGS TVE Cokemaking Survey, 2002. Cambridge, MA: China Cokemaking Team, Massachusetts Institute of Technology.

AGS MRP Team. 2001. *2001 AGS MRP Field Trip Notes.*

AGS MRP Team. 2002. *2002 AGS MRP Field Trip Notes.*

AGS MRP Team. 2004. *2004 AGS MRP Field Trip Notes.*

AGS. 1998. AGS TVE Cokemaking Survey, 1998. Cambridge, MA: China Cokemaking Team, Massachusetts Institute of Technology.

AGS. 1999. AGS SOE Cokemaking Survey, 1999. Cambridge, MA: China Cokemaking Team, Massachusetts Institute of Technology.

AGS. 2000. AGS TVE Cokemaking Survey, 2000. Cambridge, MA: China Cokemaking Team, Massachusetts Institute of Technology.

AGS. 2001. AGS SOE Cokemaking Survey, 2001. Cambridge, MA: China Cokemaking Team, Massachusetts Institute of Technology.

Akatsuka, Takeyoshi. 2001. *Modeling and Evaluation of the Transportation Sector in the Coke-Making Industry of Shanxi Province, China* (in Japanese). Master's Thesis. The University of Tokyo.

Chen, Hao. 2000. *Technological Evaluation and Policy Analysis for Cokemaking: A Case Study of Cokemaking Plants in Shanxi Province, China.* Master's Thesis. Technology and Policy Program, Massachusetts Institute of Technology, Cambridge, MA (June).

Chen, Yan. 2003. "GIS-based Planning Support System for Transportation and Industrial Location Analyses: A Case Study of the Cokemaking Sector in Shanxi Province, China." Master's thesis. Department of Urban Studies and Planning, Massachusetts Institute of Technology, Cambridge, MA (February).

China State Development and Planning Committee. 2001. *China Tenth Five-Year Plan.* Beijing: China People's Press.

DiPasquale, Denise, and William C. Wheaton. 1996. *Urban Economics and Real Estate Markets.* Englewood Cliffs, NJ: Prentice Hall, pp. 91-123

ESRI (Environmental Systems Research Institute, Inc.). 1997. *Understanding GIS, The Arc/Info Method.* New York: John Wiley & Sons.

Gittings, B. M., T. M. Sloan, R. G. Healey, S. Dowers, and T. C. Waugh. 1993. *Meeting Expectations: A View of GIS Performance Issues, Geographical Information Handling.* London: Wiley. pp. 33-45.

Kraines, Steven B., Takeyoshi Akatsuka, Larry Crissman, Karen R. Polenske, and Hiroshi Komiyama. 2003. "Pollution and Cost in the Coke-making Supply Chain in Shanxi Province, China: Applying an Integrated System Model to Siting and Transportation Tradeoffs." *Journal of Industrial Ecology* 6(3-4): 161-184.

Polenske, Karen R., and Francis C. McMichael. 2002. *A Chinese Cokemaking Process-flow Model for Energy and Environmental Analyses. Energy Policy* 30 (10), pp. 865-883.

Rees, John, and Howard A. Stafford. 1986. Theories of Regional Growth and Industrial Location: Their Relevance for Understanding High-Technology Complexes. *Technology, Regions, and Policy*. John Rees, ed.. Totowa, NJ: Rowman & Littlefield, pp. 23-50.

State Bureau of Statistics, Department of Industry and Transportation Statistics. 2001. *China Energy Statistical Yearbook (1997-1999)*. Beijing: China Statistics Press.

State Economic and Trade Commission, State Environmental Protection Agency, and (former) Ministry of Machinery. 1997. *The Catalogue of Elimination of Heavy-polluting Technologies and Equipments (the First Series)*, by Guo Jing and Mao Zi. Directive No. 367 (June 5).

Internet Sites:

U.S. Department of Transportation. Trip Model Improvement Program, accessed 06/07/05 http://tmip.tamu.edu/clearinghouse/docs/gis/metro/

[1] Associate, Morgan Stanley; At time of writing, Multiregional Planning (MRP) Research staff, MIT, Cambridge, MA, USA.

[2] Associate Professor, Department of Chemical Systems Engineering, University of Tokyo.

[3] Professor, Department of Urban Studies and Planning, MIT; Head, China Cokemaking Team, MIT, Cambridge, MA, USA.

CHAPTER 7

HUMAN EXPOSURE TO ULTRAFINE PARTICULATES IN THE COKEMAKING INDUSTRY IN SHANXI PROVINCE

QIAN Zhiqiang,[1] Uli MATTER,[2] Hans C. SIEGMANN,[3] FANG Jinghua,[4] and Karen R. POLENSKE[5]

7.0 Introduction and Scope

Shanxi Province holds the dubious distinction of being among the most polluted provinces in China. Shanxi's severe particulate air pollution, emanating mostly from coke plants, is suspected of causing a number of human ailments, ranging from eye irritation to severe diseases of the pulmonary system, including lung cancer (Pope III et al. 2002; Bömmel et al. 2000; Beckett 2000; Brüske-Hohlfeld et al. 1999). Because high air pollution levels can pose significant economic and human costs, we believe it is of importance to people everywhere to understand how and to what extent such severe particulate air pollution may affect human health and thus to be able to judge the potential hazards of their own air quality. Although in this chapter, we focus specifically on the coke industry, our conclusions are of far-reaching significance.

We designed our research on cokemaking pollution to establish a basis for exploring the physical and chemical properties of ultrafine particles so as to characterize them and evaluate their potential impact on human health. Ultrafine particles can deposit themselves deep in the human respiratory tract, where the defence mechanisms of the body are weak. Unfortunately, such small, airborne particles are particularly difficult to detect and characterize. Although researchers have established that high concentrations of small particles in the air correlate with the occurrence of cardiopulmonary diseases and mortality (Pope III et al. 2002), the specific nature of the harmful particles, especially the determination of the source of the particle and what makes one kind of particle more harmful than another, has not been clearly determined.

Karen R. Polenske (ed.), The Technology-Energy-Environmental-Health (TEEH) Chain in China: A Case Study of Cokemaking, 109–131.

In particular, we chose to focus on polycyclic aromatic hydrocarbon (PAH) molecules, which are formed during the combustion of organic material, such as coal. PAH molecules are particularly damaging to human health, causing various kinds of cancer and lung disease. Typically, PAHs are measured by collecting samples over a finite time period (for example 8 hours) on a filter and then analyzing the samples for specific PAH types and determining the total PAH mass. Unfortunately, the results of this chemical-analysis approach are made less significant by the very minute amount of material collected and the evaporation or condensation of PAH molecules during collection and extraction. In addition, this approach of speciation and mass determination of individual PAHs is complex and expensive.

The conventional technique is far too expensive and elaborate to be used in the type of field studies necessary for the scope of the present work. Thus, in our study, we have developed a new, more practical and informative technique to determine present real-time measurements of personal exposure to particle-bound PAH (PPAH) in different outdoor and indoor environments. For our study, we used the following three particulate sensors to measure the attributes of particulates, such as concentrations, active surface size, mass, and source.

First, we use portable, battery-operated sensors (based on photoelectric charging--PC) to determine PAH concentrations down to a level of 10^{-9} g per m^3 (ng/m^3) (Hart et al. 1993; Burtscher and Siegmann 1994). The devices enable us to measure human exposure in both indoor and outdoor environments in 10 to 120 seconds intervals. The short intervals helped us to detect dynamic changes of the ultrafine particles and relate these changes to surrounding activities, such as diesel trucks passing, coke being pushed, and cigarette smoking.

The second method we use to characterize particles is diffusion charging (DC), which helps us determine the size of the "active" surface of fine particulates. The surface determines the mobility and diffusion constant of the particles and is thus responsible for the degree of penetration into the human respiratory tract. Through our work, by combining the two methods of PC and DC measurement, we have discovered that the PC/DC measurement ratio is a good indicator of the source (e.g., coal combustion, cigarette smoke, diesel fumes, etc.) of emitted particles. It also helps us to track the process and time of formation of the particles in the atmospheric environment. We conduct these DC measurements in a continuous and automatic manner, similar to the PC measurements.

To be able to connect our results to the more traditional particle-measurement techniques, we also use a commercially available instrument based on light scattering (LS). LS responds to larger particles similar to the traditional methods that determine the particle concentration by collecting the particles on filters and determining their mass by weighing the filter. In addition, LS collects dynamic information on the particle, such as its process of formation.

For our research, not only do we measure coke workers' exposure to PAHs in plants, but also in their homes, so that we may establish a valid picture of their total exposure to ultrafine particles. We compare the data for particulate exposures in Chinese homes to the residential cooking exposures found in other locations.

7.1 Experimental Technique

Our purpose is to understand the physical and chemical impact of nanometric (ultrafine) particles in a complex carrier gas, such as the atmosphere, on human health.

In our experiments, we therefore focus on the surface, because with ultrafine particles most atoms or molecules are located in a thin layer at the surface and because surface science has established that the surface species may exhibit properties substantially different from the atoms or molecules inside the particle. To show the importance of the particle surface, we explain our motivation and thus the necessity to conduct field measurements.

First, the smaller the particle, the more important is the surface area. To estimate the relative importance of the surface, we assume a spherical particle, with radius R, and a surface region, with thickness ΔR. The ratio of surface-to-bulk atoms is then given by $3\Delta R/R$. For example, for a spherical particle with $R = 3$ nm (nanometers) and the surface layer with $\Delta R = 1$ nm, the ratio $3\Delta R/R = 1$ explains that 50% of the atoms or molecules are located in the surface layer. However, particles sized about 2 μm (micrometers) that is with a diameter that is 1000 times larger, only 0.3% of the atoms are surface atoms. This contrast proves the need especially with ultrafine particles to concentrate on the surface properties. Up to now, many analysts conducting official investigations focused on the mass of all particles smaller than 10 or 2.5 μm (PM_{10} or $PM_{2.5}$, respectively) thus totally ignoring the importance of the surface.

Second, the particle surface determines the dynamical properties and thus the deposition of the particles in the human body. Many ultrafine particles are formed through incomplete combustion of organic fuels and agglomerate from initial primary spherules of 5 nm diameter into huge agglomerates of more than 100 nm in size. The total surface area of these agglomerates is divided into an active and a passive part (Keller et al. 2001).

The active surface is that part of the surface on which carrier gas molecules are directly adsorbed. Hence, the active surface is directly related to particle growth, and the particle growth by adsorption of water determines the particle's lifetime in the atmosphere. For instance, hydrophobic PAH-loaded particles generated in combustion may have a very long lifetime because they do not adsorb water. Note that hydrophilic particles that do adsorb water act as condensation nuclei and may grow to a large size due to the ubiquitous presence of water. Such large particles are precipitated, hence having a short lifetime in the atmosphere. Particles that do not adsorb water remain small and can therefore be transported across continents. Very fine hydrophobic particles from combustion can indeed be found anywhere on the globe. The connection of the active surface and the dynamical properties arises as the active surface is the location where energy and momentum is transferred through collisions with molecules of the carrier gas. The transfer of energy and momentum depends on whether the particle surface is inert or reactive. If the surface is inert, the transformation results in specular reflection. If reactive, the result is diffuse reflection with a different rate of exchange of momentum between carrier gas and particle. The number and nature of collisions thus determines dynamic particle properties, such as mobility, friction, and diffusion. This, in turn, determines the location and efficiency of the precipitation of particles in filters. For example, the human respiratory tract is an efficient, yet very complex, particle-filter. The deposition of a particle in the respiratory tract entirely depends upon the mobility of the particle.

The passive surface is composed of the surface locked in the interior of the particle, or located in bays or cracks. The passive surface does not contribute to the instantaneous exchange of momentum and energy with the carrier gas, but instable species may survive longer there; thus, the passive surface contributes to particle changes on a longer time scale.

In the assessment of the health effects of particulate air pollution, it is thus extremely important to analyze the surface composition of ultrafine particles. There is no obvious way to achieve this task, because the chemical compounds of the surface changes as soon as there is a tiny variation in the composition of the carrier gas. For example, if an analyst brings the particle into a vacuum to visualize it with an electron microscope, many of the surface adsorbents evaporate and/or may even transform under the impact of the electron beam that is needed for visualization. Hence, although the electron microscope is an excellent tool to visualize nanostructures, it merely reveals the bizarre graphitic skeletons or other non-volatile nuclei of the particles. For example, the image of a mouse imaged in an electron microsope would simply show the skeleton of the mouse, nothing more.

This example shows why we felt it essential to develop a methodology suitable to analyze surface properties and evaluate particle impacts while the particle is suspended in its genuine carrier gas. Obviously, analysts find it impossible to measure all of the thousands of chemical compounds present in atmospheric particles; rather, they must confine themselves to a few groups of chemical species that are specific to the origin of the particle and important to its later evolution in the atmosphere. Therefore, our analysis concentrates on two important questions:

First, can the particle adsorb water? This determines its lifetime in the atmosphere and the deposition in the human respiratory tract, e.g., salts and acidic nuclei formed from combustion-generated SO_x and NO_x will generally adsorb water and thus grow to large liquid particles of an ideal spherical shape.

Second, was the particle generated by the combustion of organic fuels, the most prolific source of the human-made ultrafine particles, that impact human health? If so, some of the particles are carbonaceous and initially hydrophobic. The latter particles are likely to contain a number of toxic or chronically toxic substances, such as the PAHs. Due to the extraordinary stability of the benzene ring, PAHs are the only organic molecules to be formed at high temperatures in the combustion zone (Siegmann and Sattler 2000; Siegmann, Sattler and Siegmann 2002). Therefore, PAHs are a characteristic substance found extensively in combustion-generated particles. When the combustion exhaust gases cool, the heavy PAHs, i.e., those with four and more benzene rings, adsorb to the particle surface making it hydrophobic, i.e., unable to adsorb water. Once released into the atmosphere, photochemical reactions and ozone may attack the PAHs. This changes the particle surface from hydrophobic to hydrophilic. After this change, the particles can grow by adsorption of water and finally will precipitate.

In order to analyze the surface structure of the ultrafine particulates, we conducted simultaneously two measurements, *photoelectric charging* and *diffusion charging*. *Light scattering* was added as a third method to connect our data to the more traditional methods of particle quantification.

Photoelectric charging (PC) is usually used to analyze ultrafine particle surfaces in their gaseous environment to assess their structural and electronic characteristics. This method allows analysts to obtain source-specific chemical information of the particles apart from their PPAH-concentration, e.g., they can distinguish particles generated by cigarettes from the ones generated by a diesel motor (K. Siegmann, Scherrer, and H. C. Siegmann 1998). This method solves the most important, mainly political, question in particulate air pollution of who is responsible for the pollution. Furthermore, PC has been successfully employed in the measurement of desorption kinetics of large molecules from the particles (Kasper et al. 1999), in the detection of chiral particles consisting of biological material like amino acids (Kasper et al. 1999), in the investigation of carbon formation in combustion (K. Siegmann and Sattler 2000; K. Siegmann, Sattler, and H.C. Siegmann 2002), and even in the observation of electron relaxation times within the particle on ultra-fast time scales of 10^{-15} (Fierz 2001).

Our experiment was conducted as follows: suspended particles smaller than 1 μm emit photoelectrons and become positively charged when the carrier gas containing the particles is irradiated with the ultraviolet light of energy below the ionization threshold of the carrier-gas molecules, but above the photoelectric work function of the particles (Burtscher and Siegmann 1994). This is the minimum energy required to split an electron from a solid. We illustrate this technique that is suitable to measure the photoelectric charging of the particles in Figure 7.1. The air to be measured carrying the particles is pulled into a tube by a pump. Inside the tube, an excimer lamp emits monochromatic ultraviolet light, to which the gas carrying the particles is exposed. Some particles will emit a photoelectron and thereby become positively charged. An alternating-current (AC) field (not shown) removes the emitted electrons. After that, a filter catches all the particles. The charged particles will give rise to an electric current flowing from the filter to ground potential. The PC sensor measures this current that establishes the signal. The probability of charging a particle is given by the photoelectric yield, Y. At a given intensity and wavelength of the ultraviolet light, Y depends on the size of the surface area of a particle and on the nature of the chemical species in the outermost surface layers. Hence, we can analyze the surface chemistry.

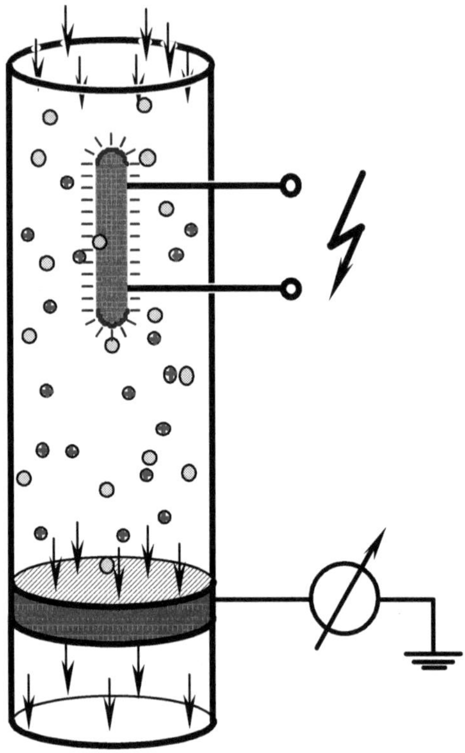

Source: The authors.

Figure 7.1: Principle of the PC Sensor Based on Photoelectric Charging

In practice, analysts have to calibrate the PC sensor to interpret its signal. The PC signal is largest with hydrophobic atmospheric particles that have adsorbed PAHs on their surface. The PC signal is correlated with the total amount of PAH bound to particles (Hart et al. 1993), because the large and flat PAH-molecules act as photonic antennas, absorbing UV-photons with great efficiency at the surface (K. Siegmann, Scherrer, and H.C. Siegmann 1998; Kasper et al. 1999; Fierz 2001; Greber et al. 1995). This causes a high photoelectric yield. Analysts have achieved the calibration of PC in terms of the density of the PPAHs by collecting the particles in a filter, extracting the PAHs, and determining their mass with gas chromatography. The results show that PPAHs can be determined from the signal of the PC-sensor in a nanogram (10^{-9} g) of PPAH/m^3 (Hart et al. 1993; Burtscher and Siegmann 1994). The calibration is very delicate mainly because of the small concentrations of PPAHs. Analysts using this measurement method may take only 1 second to do the measurement and may detect as little as 1 ng PPAH/m^3. To illustrate the sensitivity, we note that the PPAHs detected in the breath of a person can be used to determine where that person has been in the past 10 minutes, air with high PPAH concentration such as found in the street with traffic versus the clean air found perhaps on a walkway.

Diffusion Charging (DC) is the second measurement technique we apply (Keller et al. 2001). In this method, we measure the adsorption cross-section of the particle for carrier-gas ions produced in an electrical discharge. We need to use this additional method to obtain independent information on the surface of the particles, because we can use this method to analyze the active surface independently of the chemical properties.

Figure 7.2 illustrates how we can use DC to provide an electric signal that is proportional to the active surface. The device is similar to the one shown in Figure 7.1 except that the ultraviolet lamp has been replaced with a corona discharge. Applying a rather high voltage to a fine needle induces a corona discharge in the gas carrying the particles. We select the voltage on the needle and surrounding electrodes (not shown) such that the positively charged gas molecules drift slowly through the gas volume with the particles. During the time needed for a charged gas molecule to drift to the electrodes, there exists a chance that it hits a particle, and once the ion touches a surface, it will stick firmly to it due to electrostatic interaction, or else it will transfer its electric charge to the particle returning to the energetically more favorable neutral state.

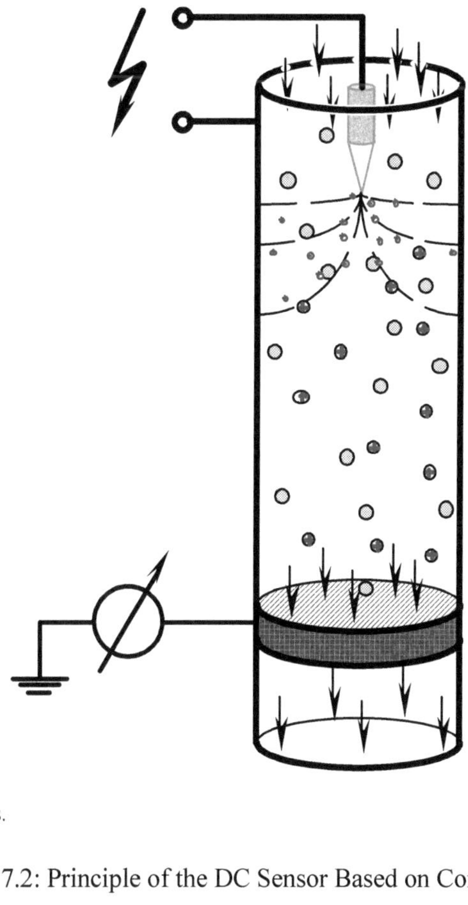

Source: The authors.

Figure 7.2: Principle of the DC Sensor Based on Corona Discharging.

In general, the probability that a gas ion hits a particle depends on the impact cross-section of the particle, which is related to the size of its active surface. By precipitating all particles in the filter and measuring the electricity that currently flows to the ground potential, an analyst obtains the charge acquired by the particles through diffusion charging. In this way, the signal of the DC-sensor is proportional to the total active surface of all particles independently of their chemical nature. To obtain valid results, analysts need to calibrate the sensor signal: to do this, they select particles of one specific active surface in a common device called a "Differential Mobility Analyzer" (DMA). The DMA selects particles of one specific "mobility diameter," which is related to the active surface through the mobility laws of particles, as shown in (Keller et al. 2001).

Analysts determine the number of particles with another common device, the condensation nucleus counter (CNC). The product of the number of particles with the individual active surfaces of each particle yields the total active surface of all the size-selected particles, which an analyst can then use to calibrate the signal of the DC-sensor in mm^2/m^3. The unit mm^2/m^3 to characterize particles is not in common use, but is strongly related to the enormous surface area of very small particles. It tells the user how much surface area in units of mm^2 is provided by the particles in a volume of 1 cubic-meter of air. Considering, first, the predominance on the surface of ultrafine particles and its significance in the transport of chemicals into the human body and, second, its importance in promoting mostly undesired hetero-chemical catalytic reactions in the atmosphere, rather than giving the total mass per cubic-meter, analysts should quote the total active surface of all particles per cubic meter. For example, consider 1 tonne of compact soot. It has a surface of $\cong 5\ m^2$. But if that same tonne of soot is dispersed in fine particles as produced by starting aircraft or by diesel motors, it has a surface of $10^8\ m^2$, enough to cover for instance the inner city of Zürich in Switzerland, allegedly a clean city, with a layer of soot. As we all know, this is not a hypothetical example—the soot layers are in plain view in all major cities, covering leaves, window sills, buildings and, invisibly, the surface of our lungs as well. We can readily estimate how much soot is produced in a certain location by noting that realistically a fraction of 0.1% of the fuel burned in a location is released into the atmosphere in the form of fine soot particles. One tonne of burned fuel produces one kilogram of fine soot particles.

Using the photoelectric-charging and the diffusion-charging sensors simultaneously, we could determine the chemical properties of the particle surface independently of the concentration of the particles. This is essential, because the ratio PC/DC is independent of particle concentration, because both PC and DC are proportional to the total surface of the particles, but by forming the ratio, this concentration-dependent total surface cancels out, and what remains is the photoelectric yield, Y, per unit particle surface. It turns out that *the ratio PC/DC depends solely on the chemical nature of the particles.* Therefore, the ratio provides a measure of the chemical composition of the active particle surface, which is characteristic for the source and mode of generation of the particles (Keller et al. 2001; K. Siegmann, Scherrer, and H.C. Siegmann 1999). More precisely, an analyst can use this result to distinguish particles that have been generated in different processes such as smoking, a diesel motor, a fire etc. In particular, if PC/DC = 0, particles have adsorbed water on their surface; this is important, as water-soluble particles, once deposited, can be disposed of more easily by the human body compared to particles that are hydrophobic and lippophilic such as PPAH, which means they dissolve in the fatty tissue and therefore have a long residence time in the human body.

Light Scattering (LS) is the third method we used in order to complement our findings on the very fine particles with information on larger particles. The light is produced by a GaAs-laser and the LS sensor measures the elastic scattering from the particles known as Rayleigh scattering into an angle of 90 degrees. The scattering increases proportional to d^6, where d is the particle diameter; hence, the signal per particle increases dramatically when the particle size increases. Therefore, as already mentioned in the introduction, the LS-sensor responds mainly to the larger particles. The intensity of the scattered light yields a rough estimate of the total mass of all particles in the PM_{10}-range because an impactor at the entrance to the LS-sensor removes particles larger than 10 μm in size. With the portable and battery-operated automatic LS-sensor, we automatically obtain an estimate of PM_{10} at the same time resolution as in the PC and DC sensors. In this way, we can relate the observations of the ultrafine particles to the appearance or presence of larger particles. Moreover, the ratio of LS/DC is roughly proportional to the particle volume, hence yields a measure of the volume of the particles that is again independent of particle concentration, but correlates with the official mass parameter PM_{10} if we can estimate the particle density.

In the field studies, we have used two portable battery-operated sensors built in the laboratory for combustion aerosols and suspended particles at the Swiss Federal Institute of Technology. These sensors measure the rate of charging of the ultrafine particles by both PC and DC. Before each field trip in China, we recalibrated the sensors in our laboratory in Switzerland and also checked for their correct functioning in the field in China. With this simple equipment, we have already investigated the ultrafine particles found in the streets of many large cities around the globe (Qian et al. 2000), as well as in many other locations and work places in Europe and the United States (P. Siegmann, K. Siegmann, and H.C. Siegmann 2002; Ott and Siegmann 2005; Velasco, P. Siegmann, and H.C. Siegmann 2005). In the present study, we also used a commercial particle detector based on scattering of light to quantify the larger particles. The use of the combination of the three particle sensors based on different physical principles is innovative, yet simple, and we note here that the equipment is capable of much further development (Fierz, Scherrer, and Burtscher 2002). The three sensors are shown in Photos 46 and 48, and the technical illustrations of the three sensors are portrayed in Photo 44.

7.2 Results: Cokemaking Plants

We used the three above-described portable monitors to conduct the measurements both in cokemaking plants and in the homes of coke workers in Shanxi Province, China, in order to evaluate particulate air pollution. Here, we present results for four types of coke ovens and examples from coke-worker homes.

Until recently, cokemaking plants in Shanxi Province operated four types of coke ovens: indigenous coke ovens, modified coke ovens, machinery coke ovens (small and large), and non-recovery coke ovens. With our measurement methods, we are able to compare pollution levels from the different coke ovens. In our study, we focus on the latter three, because at the time that we began our study in 1998, local officials had closed most indigenous coke ovens in Shanxi Province due to heavy air-pollution levels. In Chapter 3, Chen Hao and Polenske present details on the technology of these types of ovens.

7.2.1 Modified-Indigenous Coke Ovens

For modified-indigenous ovens, the cokemaking period is about two weeks, and the operators do not collect any chemical by-products.

We placed our three sensors at places close to the coke oven where coke workers work, as shown in Figure.7.3. For a typical plant, 24 coke ovens are present in each battery (refer to photo in the appendix for a portrayal of an actual set of ovens in Jiexiu, Shanxi Province.)

Also, we show in the lower part of Figure 7.4 the ratio LS/DC relating to the evolution of the average size of the particles and the ratio PC/DC, yielding the fingerprint on the surface chemistry. That day, special events include the discharge of smoke from the coke oven at 9:45 h (marked by A) and the coke-push process (i.e., unloading the coke from the oven (marked by B). First, we note that the PM_{10}-estimates are very high, ranging from 1 to 7 mg/m^3 (mg per cubic meter), which is much higher than would be acceptable in the West. Also the total active surface (reading of the DC sensor) is very high indicating high particle densities. The background level of the PC-signal, thus the PPAH concentration, is lower (100-200 ng per cubic meter) compared to the ones found in streets of European cities, but it reaches very high levels of 1,000 to 4,000 ng per cubic meter, respectively, in the special events A and B. The low level of the background of ultrafine dry particles is explained by the scavenging of the small particles by the larger ones, observed previously for instance in (Qian, et al. 2000). Yet the high levels observed in the special events of smoke from the coke oven (A) or opening the coke oven (B) shows that the modified coke ovens are very potent sources of ultrafine dry particles. The LS/DC ratio increases at event A and at event B, indicating that larger particles are present when exhaust from the coke oven is present. This corroborates with the observation of (visible!) smoke in event A and B.

The average values of PM_{10}, total active surface, and PPAH during the time period (9:00-12:30) at this working place are 1.570 mg/m^3, 1570 mm^2/m^3 and 745 ng/m^3, respectively. Given the level of general particulate pollution in the local area, the particulate pollution from the specific events during the cokemaking process (identified in the diagram as events A and B) are related exactly with the times that this type of coke oven produces heavy large particles (PM_{10}) and small particles (PPAH).

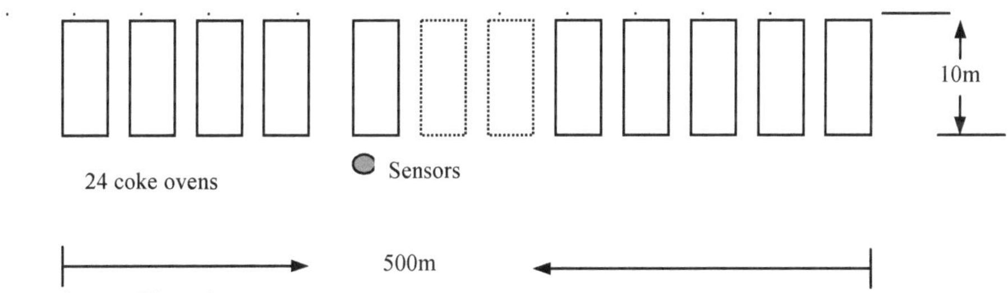

Source: The authors.

Figure 7.3: Placement of Sensors at Taiyuan Jixiehua 1975 (75-type) Coke Ovens

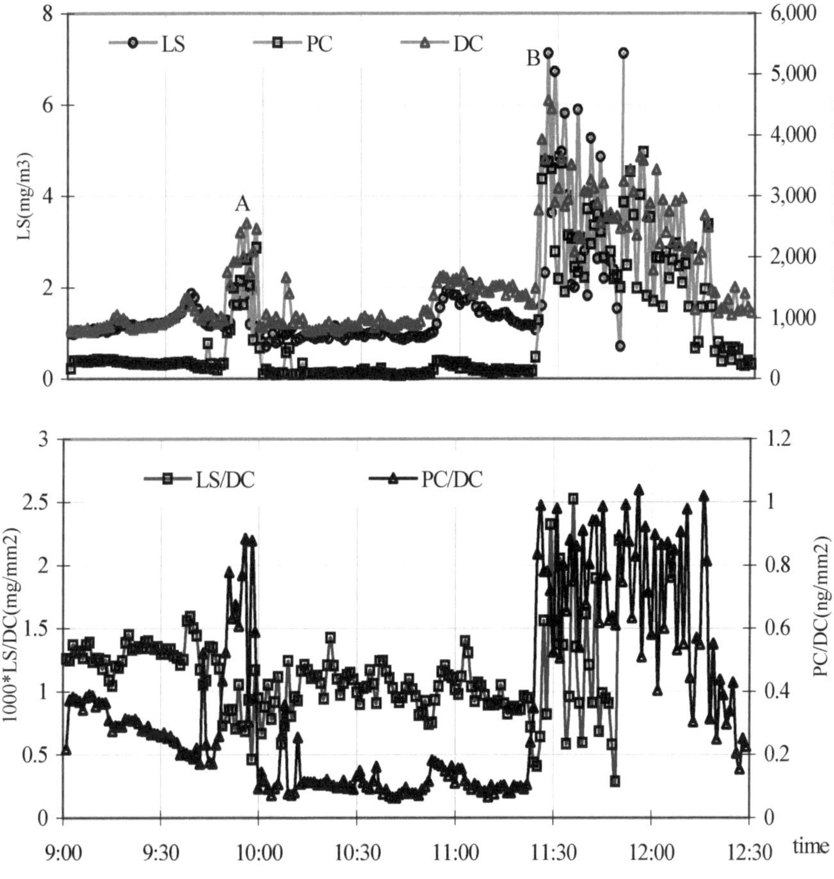

Source: Measured by Zhang Rongshen on November 17, 2001.
Note: Measurements of particulate air pollution in a coke-worker's working position at a 75-type modified oven. "A" indicates smoke from coke oven, "B" indicates coke-unloading process.

Figure 7.4: Measurements of Particulate Air Pollution at a 75-type Modified Oven

7.2.2 Small-Machinery Coke Ovens

For small-machinery coke ovens, the average cokemaking period is two days and tar, and chemical by-products, such as ammonia, are collected. We placed the sensors on the top of the coke oven where one or more coke workers operate for extended periods of the day (Photo 46).

The upper part in Figure 7.5 shows the very high PM_{10} (given by LS) values with an average of 1.480 mg/m^3. Those large particles are emitted during the cokemaking process, especially during the process of the coke push (unloading). The total active surface (given by DC) is also high with an average value of 1380 mm^2/m^3, therefore high particle densities. PC signals are as high as an average of 518 ng/m^3, and the PPAH is lower compared to those found in the street emitted by automobile traffic. Some of ultrafine particles are scavenged by large particles. Each peak in Figure 7.5 corresponds with a coke push occurring nearby. The biggest one at 10:50 is related to a coke push from an oven that is very close to the sensors.

In the lower part of the Figure 7.5, at 10:50, the PC/DC value increased dramatically, which indicates the difference in the particle sources during the coke-pushing process. The peak of LS/DC ratio at 10:50 shows large-size particles present during the coke push, with visible smoke and dust.

It is clear that the cokemaking process by this type of small-machinery coke oven produces both heavy large and small particles. We note that as of 2005, authorities in Shanxi Province are trying to close many of these small-machinery coke plants.

7.2.3 Large-Machinery Coke Ovens

We took the large-machinery coke-plant measurements at a coke-worker's working position. Average values of PM_{10}, total active surface, and PPAH are 0.332 mg/m^3, 475 mm^2/m^3, and 195 ng/m^3, respectively, at the coal-loading place for large-machinery coke ovens. The value of large particles of PM_{10} is much lower than the measurements we took at working places in other coke ovens (modified, small-machinery, and clean), although it is still higher than what is commonly acceptable. The ultrafine particle PPAH value is at the same level as for clean coke ovens and much lower than those found in modified and small-machinery coke ovens.

In the upper part of Figure 7.6, the DC/PC/LS values periodically peak every 15 minutes, which is especially clear for the DC values. The coke-quenching car was passing every 15 minutes with visible smoke and dust. The results show that the coke-loading process is the main pollution source for this large-machinery coke oven.

In the lower part of Figure 7.6, PC/DC peaks during 13:30 - 14:30 must be from fine particles; they are different from those during other times, probably because they are mostly from the diesel quenching car. LS/DC peaks indicate some large particles that could be from water vapor; therefore, they are not harmful.

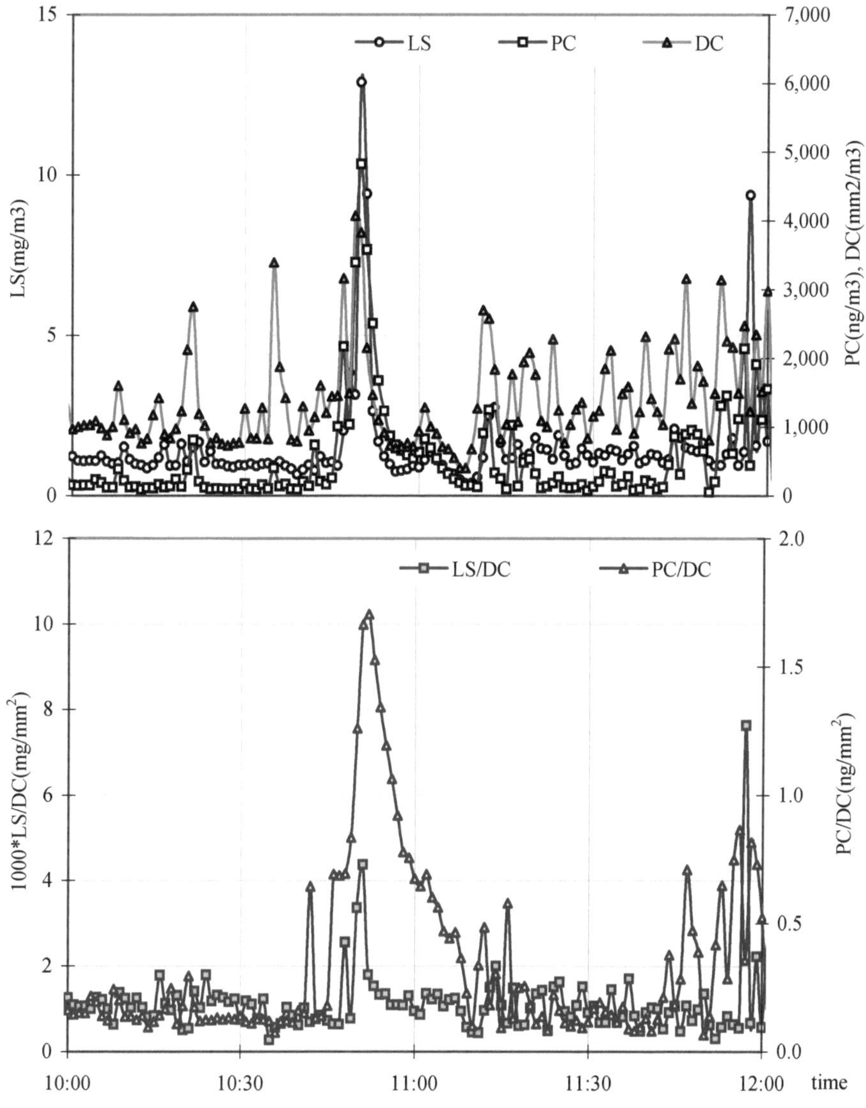

Source: Measured by Zhang Rongshen on November 20, 2001.
Note: Measurements of particulate air pollution in a coke-worker's working position at a small machinery coke oven, with peaks related to the coke-push process.

Figure 7.5: Measurements of Particulate Air Pollution at a Small Machinery Coke Oven

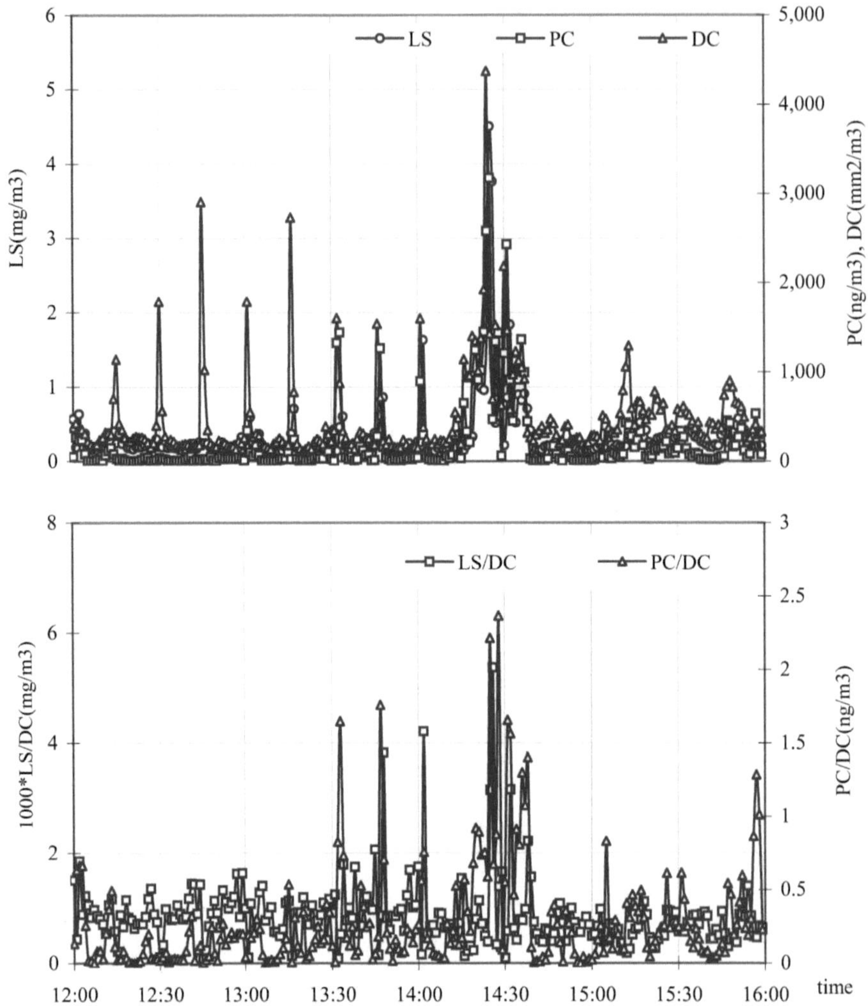

Source: Measured by Zhang Pangshen on January 2, 2003.
Note: Measurements of particulate air pollution at coke worker's working position for coal loading at a large-machinery coke oven, with a coke-quenching car passing every 15 minutes.

Figure 7.6: Measurements of Particulate Air Pollution for Coal Loading at a Large-Machinery Coke Oven

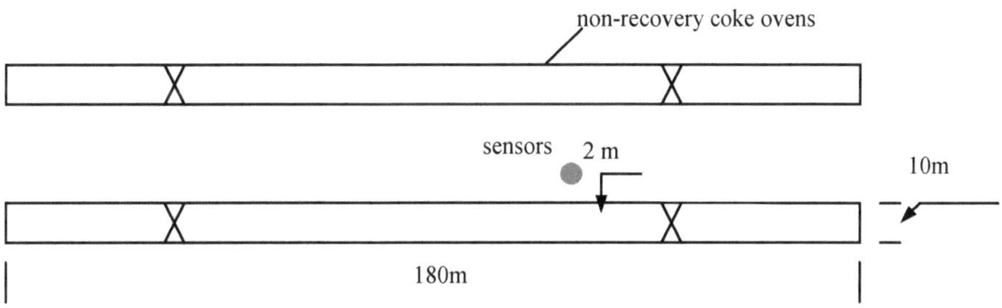

Source: The authors.

Figure 7.7: Sensor Placement on Nonrecovery (Clean) Coke Oven

7.2.4 Clean (Non-Recovery) Coke Ovens

The cokemaking period for clean coke ovens (e.g., Sanjia-96) is two weeks, and all chemical by-products are burnt and used to make power.

We placed the three sensors at a site that is about two meters away from the coke oven (Figure 7.7).

In the upper part of Figure 7.8, the PM_{10} (LS) values are in the range of 2.3mg/m^3-3.2 mg/m^3. The average is 2.675 mg/m3, showing that these types of coke ovens have very high PM_{10} values. We note, however, that there were always huge amounts of water vapor from coke-quenching nearby, thus probably the water vapor contributes a lot to the PM_{10} values, but the water vapor is harmless to humans.

High DC values show the existence of high total active surfaces, with an average of 1041 mm^2/m^3, and high particle density. PC values are in the range of 100-330 ng/m^3, with an average of 180 ng/m^3. Thus, at this monitoring station for this clean coke oven, relatively much less PPAH was emitted compared with the PPAH found in the modified and small-machinery coke ovens. This is correlated with the fact that all chemicals are burnt and used for power generation.

In the lower part of Figure 7.8, we see that the 1000*LS/DC ratio is around 0.7 mg/mm^2, and the peaks may relate to water vapor from the periodic quenching of the coke. The PC/DC ratio is in range of 0.1-0.3 ng/mm^2, similar to those during the time of no special events and during coke-loading in other coke ovens.

7.3 Results: Coke-Worker Homes

In coke-worker homes, the measured results show typical indoor air-pollution levels for China.

Although we expected to find that human exposure to particles at the coke plants also occurs in the homes of coke workers, until now, such data have been lacking. So far, we

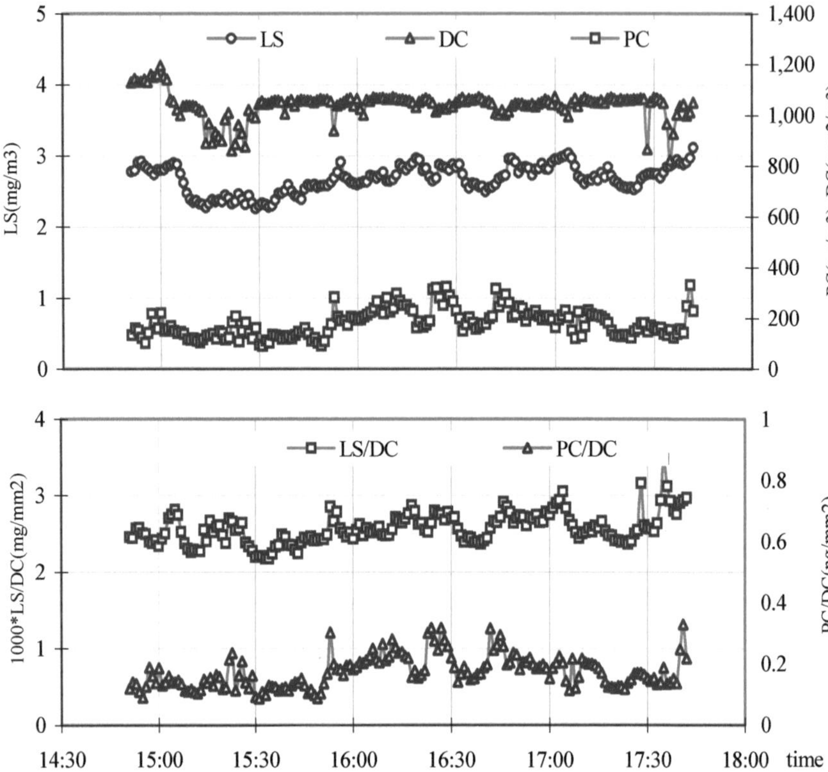

Source: Measured by Qian Zhiqiang on December 28, 2002.
Note: Measurements of particulate air pollution in a coke-worker's working position for a fire observer at a nonrecovery coke oven, huge amounts of water vapor from coke quenching nearby.

Figure 7.8: Measurements of Particulate Air Pollution for a Fire Observer at a Clean Coke Oven

have found that heavy particle emissions occur at the working places in cokemaking plants. Now, we present similar results for particulate air pollution in the homes of coke workers and demonstrate the sources of particles related with common activities of everyday life.

The typical coke-worker home in towns and villages in Shanxi Province includes a visiting room with one coal stove for cooking (and heating during winter season), several bedrooms, and an additional open field, in some cases. No home has air ventilation, and each uses coal for the coal stove, which features a channel for flue gas to the outside.

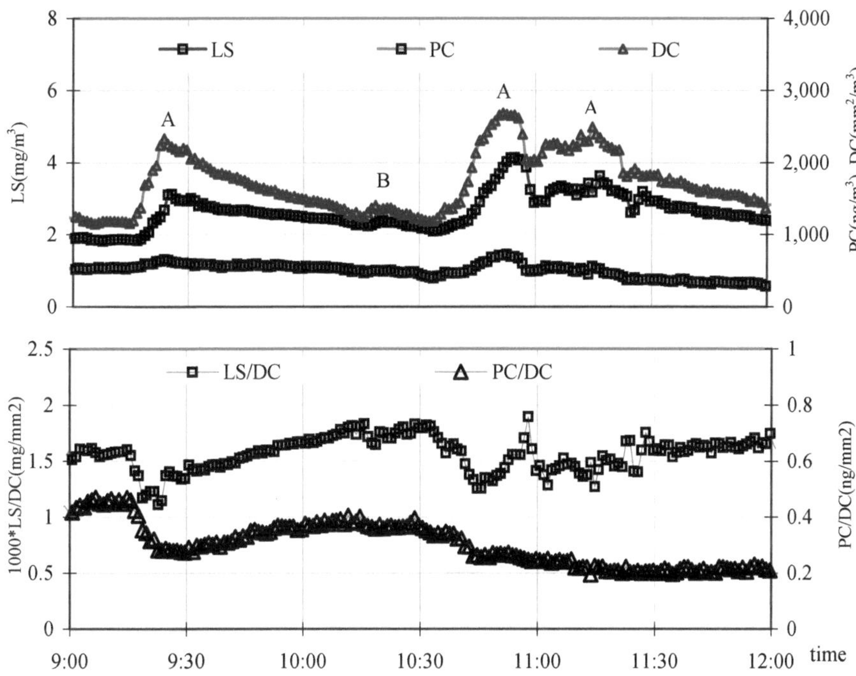

Source: Measured by Tan Wenyuan on December 29, 2000.
Note: Measurements of particulate air pollution in a coke-worker's home, living and visiting room with coal stove for hot water and heating.
Peak "A" is related to cigarettes smoking, and peak "B" is related with coal being loaded into a coal stove for hot water and cooking.

Figure 7.9: Measurements of Particulate Air Pollution in a Coke-Worker's Home with a Coal Stove

Figure 7.9 shows the measurements of particulate air pollution in a visiting room of about 15 m^2 in the home of a coke worker. In the upper part of the figure, peaks marked by "A" are the highest values for DC/PC/LS, respectively, and we can relate most peaks to the occurrence of cigarette smoking. The peak marked B is related with a person loading coal into the coal stove, which was being used both for heating and cooking.

The average values of LS/PC/DC in this morning are: PM_{10}, 2.652mg/m^3, PPAH, 501 ng/m^3, and total active surface 1,722 mm^2/m^3. Therefore, we measure a high pollution of both large particles and small ones due to the cigarette smoking and use of a coal stove inside the room.

In the lower part of Figure 7.9, cigarette smoking, coal burning, and coal loading are the visible sources for particulate pollution in this closed-room. This is correlated with PC/DC ratio values increasing from 0.3 ng/m^2 to 0.4 ng/m^2 at the coal-loading time of 10:20, and the PC/DC ratio remained at around 0.2--0.3 ng/m^2 while cigarettes were smoked, with the peak marked "A" in the upper part of the figure; LS/DC indicates that there were large particles in this room during the entire morning.

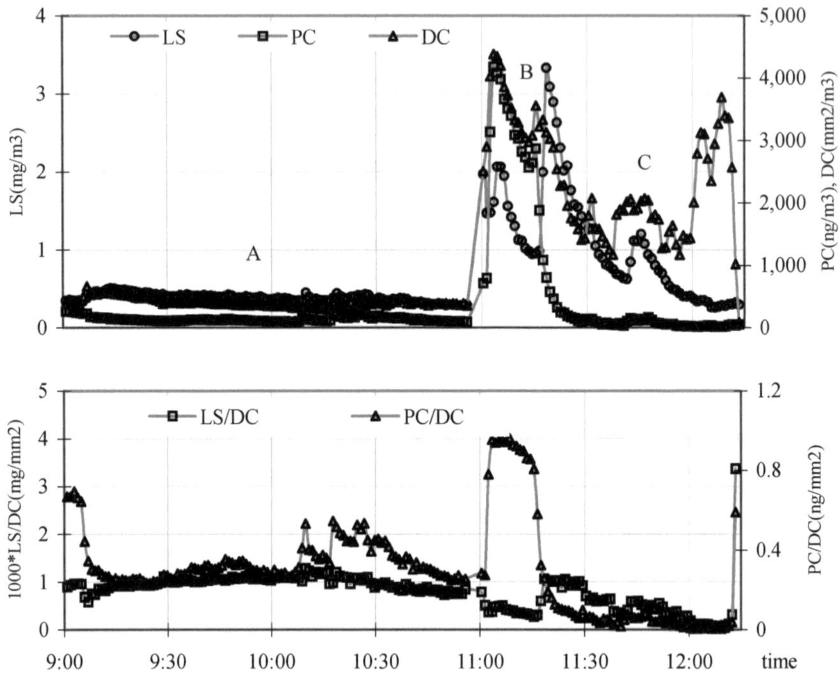

S

Source: Measured by Zhang Pangshen on January 1, 2001.
Note: Measurements of particulate air pollution in a coke-worker's home, living and visiting room with coal stove for hot water and heating.
"A" represents 9:00-11:00 living and visiting room, "B" represents 11:00-12:00 cooking room, and "C" represents cigarette smoking.

Figure 7.10 Measurements of Particulate Air Pollution in a Coke-Worker's Home with a Coal Stove

Figure 7.10 shows the measurements of particulate air pollution in another coke worker's home. The combined area of the living and visiting rooms is about 20 m^2. These rooms feature a coal stove for heating. The cooking room is about 6 m^2.

We took the measurements in the living and visiting room from 9:00-11:00, and in the cooking room from 11:00-12:00; during this period, a woman was cooking the meals. In the upper part of Figure 7.10, from 9:00-11:00 in the visiting and living room, the LS. DC, PC values were stable, because there were no activities except that the family was using the coal stove for heat. The average PM_{10}, total active surface and PPAH values are 0.390 mg/m^3, 406 mm^2/m^3, and 138 ng/m^3 respectively. Even such comparatively low values of PPAH supersede the levels found in homes in California by more than an order of magnitude (Ott and Siegmann 2005).

In the lower part of Figure 7.10, some peaks occur for PC/DC from 9:00-11:00, which may be related with putting coal into the stove, whereas, the relatively stable LS/DC ratio indicates that no large particles were being emitted.

However, particulate concentrations in the cooking room during 11:00-12:00 were high. In the upper part of Figure 7.10, we see that the LS, DC, and PC values increased during the cooking time. Peaks marked by B occurred exactly when cooking was being done; the peak marked by C is related with cigarette smoking by the cook. The average values of PM_{10}, active surface, and PPAH are 1.08 mg/m^3, 2361 mm^2/m^{3}m, and 761 ng/m^3, respectively. We find high values of PM_{10}, i.e., large-size particles, dense particles, and concentrated small particles (PPAH) were produced heavily during this cooking time because the typical Chinese cooking involves oil, and the cook was smoking a cigarette. Heating oil can produce large amounts of PPAH, where the PAHs are not generated in the combustion, but evaporate from the heated oil (Siegmann and Sattler 1996). In the lower part of Figure 7.10, at 11:00, the PC/DC ratio peak shows another dangerous ultrafine particle source, that is, the PPAH from cooking.

Now, we present another case in Figure 7.11 of particulate air pollution in a coke-worker's home, which is a number of kilometres away from the coke plants and located near a town. There is no coal stove for heating in the visiting room, and the only visible human activity is cigarette smoking. The outside door to the house is open.

In the upper part of Figure 7.11, the peak marked by A is related with cigarette smoking. The average values of PM_{10}, active surface, and PPAH are: 1.514 mg/m^3, 906 mm^2/m^3, and 473 ng/m^3, respectively. We found high values of PM_{10}, that is, large-size particles, dense particle numbers for the DC sensor values, and harmful levels of ultrafine particles PPAH for the PC sensor.

In the lower part of Figure 7.11, the peaks of LS/DC and PC/DC correspond to Peak A in the upper part of the figure, indicating that both large and small particles occur.
The indoor air pollution is heavy again, yet local people live with this poor air quality everyday.

Finally, we show a case of particulate air pollution in the open field of a coke-worker's home and a bedroom overnight to determine the air quality outside the home and in the bedroom, with a coal stove again being used for heating. We conduct these measurements in order to determine particle sources from sources other than the home itself and how much people are exposed to particles when sleeping.

In Figure 7.12 from 12:35 to 17:05, we conduct measurements in an open field outside the house; therefore, the values more or less show us the pollution level in the ambient air in the local area generally. The average values of PM_{10}, active surface, and PPAH during the measurement period are: 1.05 mg/m^3, 546 mm^2/m^3, and 112 ng/m^3, respectively. The large PM_{10} value means a high level of large-size particles from local industries, such as coke plants, dust, and smoke from the heating and/or cooking from each household in the village. The PPAH level is reasonably low for there is no nearby visible combustion.

The average values of PM_{10}, active surface, and PPAH in the bed room from 17:05 to 24:00 were: 1.99 mg/m^3, 1058 mm^2/m^3, and 643 ng/m^3, respectively. We found obvious heavy particulate air pollution (both large and ultrafine).

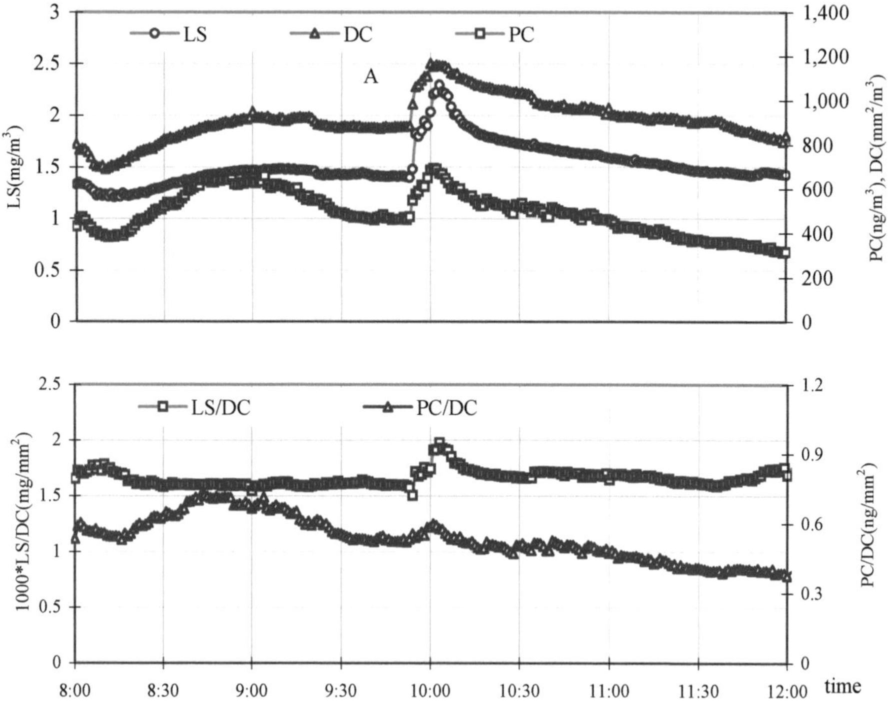

Source: Measured by Qian Zhiqiang on December 30, 2002.
Note: Measurements of particulate air pollution in a coke-worker's home: living room.
"A" represents 9:54 cigarette smoking, door to outside is open

Figure 7.11: Measurements of Particulate Air Pollution in a Coke-Worker's Home: Living Room

The peaks in Figure 7.12 are related with events, possibly from cigarette smoking and a coal stove.

7.4 Conclusion

In conclusion, our measurements of particulate air pollution in the coke plants and in the homes of coke-workers demonstrate the following:

1. Modified-indigenous, small-machinery, and clean coke ovens produce very heavy PM10. However, high PM10 values in the clean coke plant may come from the water vapor that occurs when coke is quenched with water. The former particulates are very harmful to humans, but the latter particulates are harmless.
2. Modified-indigenous and small-machinery coke ovens also produce severe concentrations of PPAH.

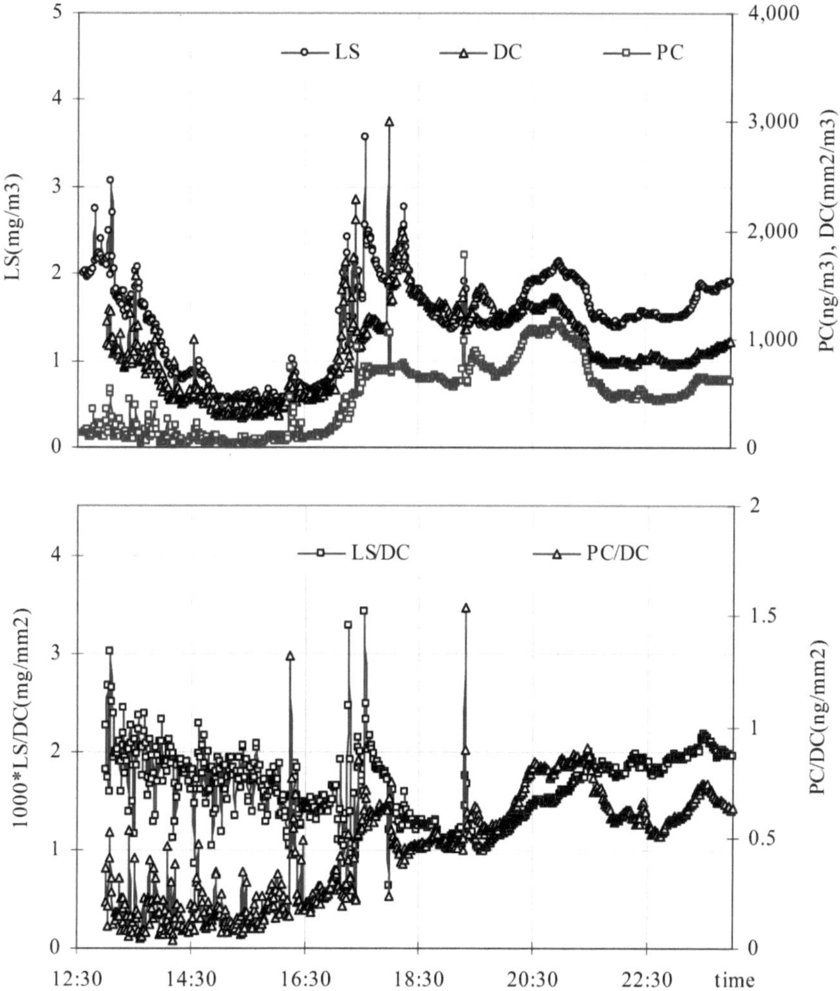

Source: Measured by Zhang Rongshen on November 21, 2001.
Note: Measurements of particulate air pollution in a coke-worker's home. 12:35-17:05 is in an open field outside the home, 17:05-24:00 is in the bedroom.

Figure 7.12: Measurements of Particulate Air Pollution in a Coke-Worker's Home: Outside Home and Inside Bedroom

3. Large-machinery coke ovens emit low levels of PM_{10} compared with other types of coke ovens; also large-machinery and clean coke ovens emit relatively low levels of PPAH. However, at another working place at the clean coke oven, we find high PPAH values; therefore, the particulate air-pollution level relates not only to the type of coke oven, but also to the working position of the workers.

4. Particulate air pollution (both PM_{10} and PPAH) at the coke-worker's home is heavy. Cigarette smoking, coal stoves, and the Chinese-method of cooking with oil are the main sources.

In Chapter 9, Polenske discusses some of the future possibilities for conducting measurements of the health effects on coke workers and people in the communities near to the coke plants.

References

Andreas, M.O. 2001. The Dark Side of Aerosols. *Nature*. **409**: 671-672.

Beckett, W.S. 2000. Occupational Respiratory Diseases. *The New England Journal of Medicine*. **342**(6): 406-413.

Bömmel, H., M. Li-Weber, E. Serfling, and A. Duschl. 2000. The Environmental Pollutant Pyrene Induces the Production of IL-4. *Journal of Allergy and Clinical Immunology*. **105**(4): 796-802.

Brüske-Hohlfeld, I., M. Möhner, W. Ahrens, H. Pohlabeln, J. Heinrich, M. Kreuzer, K. Jöckel, and H. Wichmann. 1999. Lung Cancer Risk in Male Workers Occupationally Exposed to Diesel Motor Emissions in Germany. *American Journal of Industrial Medicine*. **36**(4): 405-414.

Burtscher, H., and H.C. Siegmann. 1994. Monitoring PAH Emissions from Combustion Processes by Photoelectric. *Combustion Science and Technology*. **101**: 327-332.

Chuang, J.C., P.G. Callahan, C.W. Lyu, and N.K. Wilson. 1999. Polycyclic aromatic hydrocarbon exposures of children in low-income families. *Journal of Exposure Analysis and Environmental Epidemiology*. **9**(2): 85-98.

Denissenko, M.F., A. Pao, M. Tang, and G.P. Pfeifer. 1996. Preferential Formation of Benzo[a]pyrene Adducts at Lung Cancer Mutational Hotspots in P53. *Science*. **274**(5286): 430-432.

Fierz, M., 2001. Electronic Relaxation in Metallic Nanoparticles. *Physics of Low Dimensional Systems*. J.L. Moran-Lopez. Ed. New York: Plenum Publishers: 57-66.

Fierz, M., L. Scherrer, and H. Burtscher. Real-time Measurement of Aerosol Size Distributions with an Electrical Diffusion Battery. *Journal of Aerosol Science*. **33**(7), 1049-1060 (2002)

Greber, T., T. Gießel, C. Pettenkofer, H.C. Siegmann, and G. Ertl. 1995. Substrate Mediated Autoionization of Benzene on Graphite. *Surface Science*. **343**(3): L 1187–L1191.

Hart, K.M., St. R. McDow, W. Giger, D. Steiner, and H. Burtscher. 1993. The Correlation between In-Situ, Real-Time Aerosol Photoemission Intensity and Particulate Polycyclic Aromatic Hydrocarbon Concentration in Combustion Aerosols. *Water, Air and Soil Pollution*. 68, 75-90. Kluwer Academic.

Kasper, M., A. Keller, J. Paul, K. Siegmann, and H.C. Siegmann. 1999. Photoelectron Spectroscopy without Vacuum: Nanoparticles in Gas Suspension. *Journal of Electron Spectroscopy and Related Phenomena*. **98-99**: 83-89.

Keller, A., M. Fierz, K. Siegmann, H.C. Siegmann, and A. Filippov. 2001. Surface Science with Nanosized Particles in a Carrier Gas. *Journal of Vacuum Science & Technology A*. **19**(1): 1-8.

Ott, W.R., and H.C. Siegmann. 2005. Using Multiple Continuous Fine Particle Monitors to Evaluate Tobacco, Incense, Cooking, Wood Burning, and Vehicular Sources in Indoor, Outdoor, and In-Transit Settings. To be published.

Ott, W.R., and J.W. Roberts, 1998. Everyday Exposure to Toxic Pollutants. *Scientific American*. **278**(2): 86-91.

Peters, A., J. Skorkovsky, F. Kotesovec, J. Brynda, C. Spix, H. E. Wichmann, and J. Heinrich. 2000. Associations between Mortality and Air Pollution in Central Europe. *Environmental Health Perspectives*. **108**(4): 283-287.

Pope III, C. A., R.T. Burnett, M.J. Thun, E.E. Calle, D. Krewski, K. Ito, and G.D. Thurston. 2002. Lung Cancer, Cardiopulmonary Mortality, and Long-term Exposure to Fine Particulate Air Pollution. *Journal of the American Medical Association (JAMA)*. **287**(9): 1132-1140.

Qian, Z., K. Siegmann, A. Keller, U. Matter, L. Scherrer and H. C. Siegmann. 2000. Nanoparticle Air Pollution in Major Cities and Its Origin. *Atmospheric Environment*. **34**: 443-451.

K. Siegmann and K. Sattler. 1996. Aerosol from Hot Cooking Oil: A Possible Health Hazard. *Journal of Aerosol Science*. **1**: 493-494.

Siegmann, K. and K. Sattler, 2000. Formation Mechanism for Polycyclic Aromatic Hydrocarbons in Methane Flames. *Journal of Chemical Physics*. **112**(2): 698-709.

Siegmann, K., K. Sattler, H.C. Siegmann. 2002. Clustering at High Temperatures: Carbon Formation in Combustion. *Journal of Electron Spectroscopy and Related Phenomena*. **126**(1-3): 191-202.

Siegmann, K., L. Scherrer, and H.C. Siegmann. 1998. Physical and Chemical Properties of Airborne Nanoscale Particles and How to Measure the Impact on Human Health. *Journal of Molecular Structure: Theochem*. **458**(1-2): 191-201.

Siegmann, Ph., K. Siegmann, and H.C. Siegmann. 2002. Polución de Nanoparticulas en el Aire de las Principales Vias de Madrid. *Revista Espaniola de Fisica*. **16**(4): 24-28.

Velasco, E., P. Siegmann, and H.C. Siegmann. 2004. Exploratory Study of Nanoparticles Carrying Particle-bound Polycyclic Aromatic Hydrocarbons in Different Environments of Mexico City. *Atmospheric Environment* **38**: 4957-4968.

---

[1] Research Associate, Laboratory of Suspended Particles and Combustion Aerosols, Swiss Federal Institute of Technology. Zürich.

[2] Senior Researcher, Matter Engineering AG, Nanoparticle Measurement, CH 5610 Wohlen, Switzerland.

[3] Professor of Physics, Stanford Linear Accelerator Center, Stanford University, Stanford Ca 94309, USA

[4] Professor of Chemical Engineering, Taiyuan University of Technology, Taiyuan, China.

[5] Professor, Department of Urban Studies and Planning, MIT; and Head, China Cokemaking Team, Cambridge, MA, USA.

CHAPTER 8

A SOCIOECONOMIC PERSPECTIVE ON COKEMAKING IN SHANXI PROVINCE

Hoi-Yan Erica CHAN[1] and Holly KRAMBECK[2]

8.0 Introduction

In addition to having profound economic and environmental implications for the region, the rise of the cokemaking sector in Shanxi Province, China, directly affects the livelihoods and health of residents and workers. Over the past two decades, cokemaking Township and Village Enterprises (TVEs) have become increasingly important in generating provincial income and employment. Currently, TVEs produce 85–90% of the province's total coke output (Chapter 2). In this chapter, we examine the socioeconomic implications of the booming TVE cokemaking industry in Shanxi Province, focusing on the labor dimensions. We look at how the regional economic climate, environmental policies, and technological decisions have come together to affect the region's residents and workers in fundamental areas, such as health and safety, wages, and education. In doing so, we also juxtapose cokemaking TVEs with traditional cokemaking State-Owned Enterprises (SOEs) to show how industry-wide ownership structure reforms in China are changing labor practices in the cokemaking sector.

8.1 Data and Methodology

To examine the socioeconomic implications of the Shanxi cokemaking sector, we have drawn upon a number of different sources. We use data collected from the Alliance for Global Sustainability (AGS) China Cokemaking SOE and TVE surveys (1998-2001) to examine the business and labor practices within this sector (see Chapter 2). These surveys contain questions on the enterprises' employment practices, ownership structure, and supply-chain relationships. Members of the AGS China Cokemaking

Karen R. Polenske (ed.), The Technology-Energy-Environmental-Health (TEEH) Chain in China: A Case Study of Cokemaking, 133–147.

Team administered the four TVE surveys in 1998, 2000, 2002, and 2003, sampling 158, 164, 127, and 110 plants, respectively (31 of which are common to both the 1998 and 2000 surveys).

The matched sample of 31 common plants is particularly helpful for looking at the changes and evolution of practices within cokemaking enterprises over the course of three years. We use data from those matched plants throughout this chapter to show how the cokemaking sector is changing, especially in its labor relationships, although we realize that the comparison would be improved if we could have matched plants for the 2002 and 2003 surveys as well. Team members also conducted two SOE surveys in 1999 and 2001, with information on eight plants gathered in 1999, and 49 in 2001.

Finally, we visited six cokemaking enterprises (under various ownership structures) in Shanxi Province in the summer of 2001 and another seven in January of 2004 while on field trips with the AGS China Cokemaking Team. We observed plant operations and met with managers and workers. The AGS team interviewed plant managers regarding their plants' history and evolution, business outlook, technology choice, and health and safety/employment practices. To compare China's SOEs and TVEs, and the overall socioeconomic and regional differences between Shanxi Province and other parts of China, we consult statistical yearbooks published by state and regional statistical authorities in China from the mid-1990s to 2001. These statistics provide an overall picture of the nature, structure, and institutional evolution of the TVEs.

8.2 Enterprise Ownership Structure: Background

Only in the 1950s, just after the People's Republic of China was founded, did China truly begin to industrialize. With help from Russian neighbors, SOEs were erected throughout the country, driving economic growth and providing all types of social services for their workers. Although these enterprises created a foundation for economic development in China, some of them are often criticized for being both highly polluting and on the brink of insolvency. Thus, as China transitions to a market economy, it is not surprising that many of these firms are being shut down. Today, SOEs represent only 50% of fixed-asset investment in China, although, in Shanxi Province, which is economically behind its southern neighbors, SOEs represent 63% of total fixed-asset investment (*China Statistical Yearbook* 2001).

TVEs grew out of the commune system established in the 1950s. In communes, land ownership was shared among local villagers, and production was governed and managed by Party-selected officials. Revenues from land were then, in theory, reinvested in the community. After the national government lifted some land-use restrictions in the late 1970s, many communes chose to build factories on their land, since manufacturing tended to be more profitable than farming (Chan 1999).

TVEs have consistently been more profitable than state-owned enterprises, and they have accounted for an increasing share of the nation's total output (Mahdavi 2001). TVEs' share of gross industrial output rose from 10% in 1980 to 45% in 1993; by 1993, together with urban industrial "collectives" and fully private firms, they accounted for a majority (57%) of China's industrial production (Cable 1996). In 1998, TVE output accounted for 28% of China's gross domestic product (GDP) (*ChinaOnline* 2000). Similarly, in Shanxi Province, TVEs have been a growing force, providing employment for thousands of rural residents and sparking economic development.

An important distinction between SOEs and TVEs is that TVEs are under no obligation to provide the same level of social-welfare benefits to their workers as SOEs. As a result, TVEs spend less than SOEs on worker benefits, such as medical care and housing. To illustrate, in 1999, SOEs, on average, spent 490 Renminbi (RMB) per year on social-welfare, compared with 149 RMB per year by TVEs (Mahdavi 2001). This single distinction between SOEs and TVEs has many significant socioeconomic implications for Shanxi Province, many of which shall be discussed in the following section.

8.3 Socioeconomic Impacts of Enterprise Ownership Structure

At the beginning of the 1990s, SOE coke production represented close to 80% of total coke production in Shanxi Province and TVE coke production 20%. By the early 2000s, just the reverse was true, as TVE production boomed. In this section, we discuss how changing enterprise ownership structures in Shanxi Province have affected employment patterns, educational attainment, and general livelihood of the work force.

8.3.1 Employment

Before discussing the impact of enterprise ownership structure on employment in Shanxi Province, we first review existing employment patterns. Shanxi is a relatively poor and predominantly rural region in China. In 2000, 73% of the provincial population lived in rural areas—a decrease from 90% in the 1950s. Also, the level of female labor-force participation in Shanxi Province is significantly below the national average--the female labor-force participation rate (staff and workers) was 14% in Shanxi Province in 2000 compared to 35% in China. (*China Statistical Yearbook* 2001)

Increased domestic and global demand for coke has sparked a proliferation of TVEs across Shanxi Province, providing farmers with more opportunities for industrial employment. One reason why TVEs have sprung up in greater numbers than SOEs is that TVEs, unconstrained by the national bureaucratic requirements that bind SOEs, can build, expand, and modify coke ovens more quickly. Cokemaking TVEs tend to be very labor intensive, and hence the proliferation of TVEs has had significant impacts on rural employment. Although rural TVEs provided 18% of all jobs in China in 2000 (712 million total), they accounted for only 12% of total investment in fixed assets (*China Statistical Yearbook* 2001).

Partially because of an input-mix that is labor-intensive with low capital investment, TVEs have lower labor productivity than SOEs. This means that TVEs are capable of generating a substantial amount of employment in China's countryside at lower costs than state-owned firms.

As long as TVEs continue to have a competitive advantage over SOEs by relying on cheap rural labor, TVEs will be more effective in generating rural industrial employment than SOEs. This situation may not last for long, however. Also, we note that TVEs in Shanxi Province have average wages that are higher than in the nation (Figure 8.1) As TVEs generate more revenue to invest in capital improvements, they will rely less and less on unskilled rural labor.

We have incomplete information for employment, number of coke plants, and output for SOEs versus TVEs. We do know, however, that overall, the total cokemaking employment in the five surveyed regions in Shanxi Province declined from 96,000 employees in 1995 to 75,000 in 2001, and that the TVE labor productivity increased from 350 tonnes/person in 1998 to 400 tonnes/person in 2003. The number of TVE plants declined from 1,872 in 1998 to 589 in 2003 (Chapter 2, Figure 2.4; Table 2.1, respectively). Although the average number of employees at each TVE is increasing, the total number of employees working in Shanxi cokemaking TVEs has decreased in recent years (implying that it is the less labor-efficient TVEs that are closing–this notion is supported by technology trends described in Chapters 2 and 3). Meanwhile, there has been a slight increase in the number of cokemaking SOE employees.

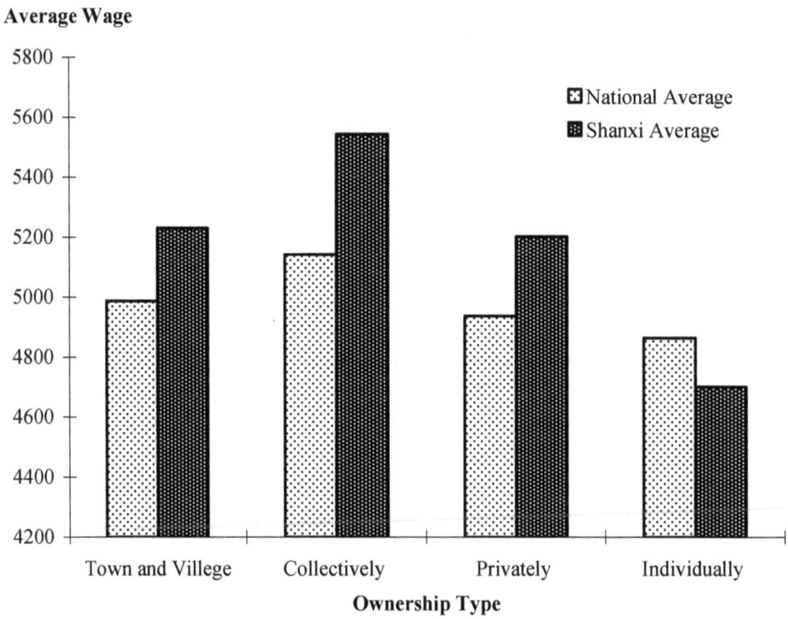

Source: Chinese Township and Village Enterprises Yearbook (1999).
Note: Wages are measured in 1998 (nominal) Renminbi.

Figure 8.1: Average Wages of TVE Employees in Shanxi Province and China, 1998

8.3.2 Education and Work Experience

The average education level of the population in Shanxi Province is below the national mean. Staff and workers employed at Shanxi's TVEs, however, have educational attainments that are significantly higher than the general population in the province and are comparable with the national averages among enterprise staff and workers in other Chinese provinces (*China Township and Village Enterprise Yearbook 1998-2002*).

The composition of the workforce is relatively uniform across time and across enterprise ownership types. In both the 1998 and 2000 surveys, Shanxi cokemaking TVEs had the following employee breakdown: 78% production workers, 8% technical staff, 11% administrative staff, and 2% other. This breakdown closely matches that of Shanxi cokemaking SOEs: 78%, 9%, 7%, and 5% for the same categories of workers respectively.

Production workers are responsible for the majority of the operations within a cokemaking facility. Their range of responsibilities include: loading and unloading coal and coke, driving the coke-quenching car, operating oven doors, and handling and transporting the final coke products.

Technical workers are responsible for the functioning of plant equipment. A typical member of the technical staff sits in front of meters and machines for eight hours a day, with five-minute breaks every hour, to ensure that the readings are normal.

Administrative staff include plant managers and office workers. In both SOEs and TVEs, on average, the administrative staff has the highest educational attainment, followed by technical, and then production, staff.

None of the production workers surveyed in 1998 and only 1% of them in 2000 had more than 12 years of schooling. The overall educational attainment among the technical staff employed in TVEs fell from 1998 to 2000. This coincided with an increase in the proportion of workers at the lower end of the educational attainment spectrum (6 years or less). Unlike the technical staff, administrative staff surveyed in the TVEs had an overall improvement in educational attainment between 1998 and 2000. A higher proportion of them received 12 to 18 years of schooling. SOE workers in all three categories had higher educational attainment (Figure 8.2).

In 1998, 10% of TVE production workers received no training for their jobs. At the same time, a majority of them received one week of training. According to AGS team member Professor Fang (2002b) from the Taiyuan University of Technology in Shanxi Province, "training" usually refers to courses offered by companies to employees on administrative and technical knowledge for different working positions. Such training is offered because the new workers are typically from farms and have little prior education. Some companies offer other new or advanced courses to older workers or office workers so that they can take on new responsibilities. Some coke plants have had their labor-force participate in classes on environmental regulations and measurements, including the ISO (International Organization for Standardization)

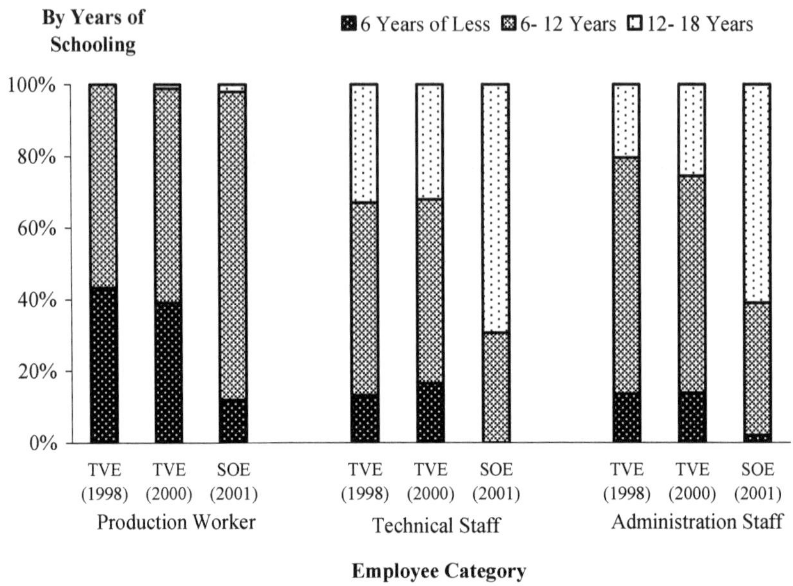

Source: AGS (1998, 2000 and 2001).
Note: The percentages represent the proportion of surveyed plants that indicated their staff received the depicted years of schooling.

Figure 8.2: Employee Educational Attainment, Shanxi Province
TVE 1998 and 2000 and SOE 2001

production standards, especially ISO 9000 and ISO 14000, the latter pertaining specifically to environmental standards.

When our cokemaking team administered the TVE 2000 survey, only 4% of TVE production workers reported having received no training. At the same time, employers seemed to have substituted comprehensive training programs with shorter, one-day training curricula. Only 44% of the production employees in 2000 received one week or more training, compared to 78% in 1998. To compare, 90% of SOE production workers received training that lasted for one week or longer. In addition, the majority of SOE technical and administrative staff received at least one month of training (Figure 8.3).

When cokemaking TVEs first came into existence, almost all entrepreneurs and workers were first-generation industrial workers, who had mostly worked in the agricultural sector. In 1998, 18% of the TVE production workers were new (had worked for one year or less). This dropped to 13% by 2000. The majority of the workers in all three job categories had worked in cokemaking for one to five years by 1998, and this trend continued through 2000, except that the majority of the administrative staff had five or more years of cokemaking experience by 2000 (Table 8.1). On the one hand, this is not surprising given the managerial and supervisory nature of the administrative positions. On the other hand, cokemaking SOEs in our surveys consist primarily of experienced workers with five or more years of experience. This difference is also not surprising, considering that the cokemaking TVEs are a relatively new sector in Shanxi Province.

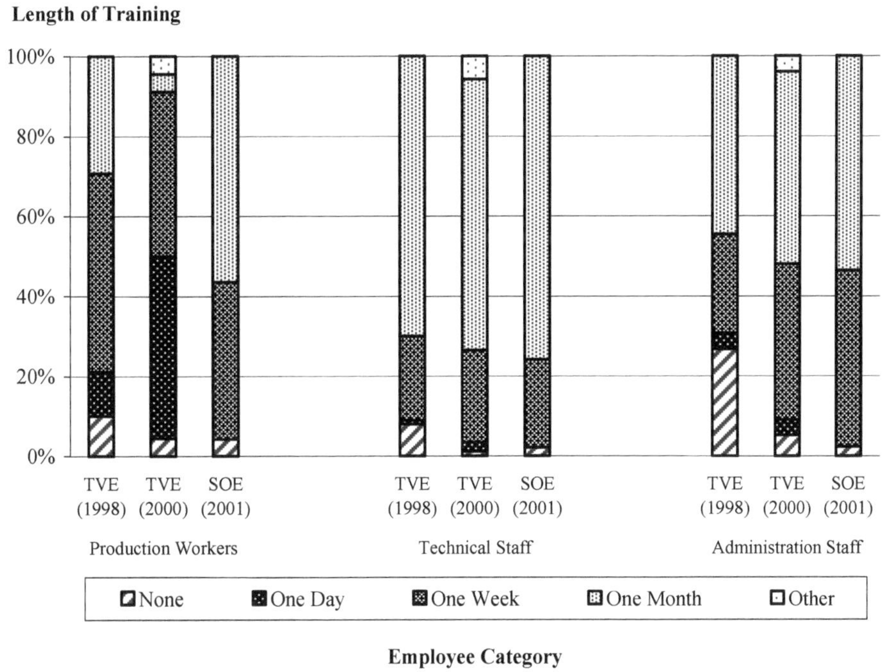

Source: AGS (1998, 2000 and 2001).

Note: The percentages represent the proportion of plants surveyed that indicated their staff received the depicted length of training.

Figure 8.3: Employee Training, TVE 1998 and 2000 and SOE 2001

Table 8.1: Shanxi Province Coke Employees' Tenure, 1998, 2000, 2001

Unit: percent

		1 year or less	1-5 Years	5 or more years
TVE-1998	Production Workers	18	57	23
	Technical Staff	6	51	37
	Administrative Staff	6	51	36
TVE-2000	Production Workers	13	46	29
	Technical Staff	5	41	40
	Administrative Staff	3	36	46
SOE-2001	Production Workers	10	37	25
	Technical Staff	6	35	45
	Administrative Staff	6	29	49

Source: AGS (1998, 2000, and 2001)

Note: The percentages represent the proportion of surveyed plants that indicated their staff had worked in the industry for the depicted length of time.

Thus, the lower educational attainment among technical workers, administrative workers and managers at the TVEs compared to SOEs implies that TVE workers without many years of schooling can leave farming activities to become industrial managers. At the same time, we note that the increase in educational attainment among TVE administrative staff, but not the other job categories, leads to a widening gap in education between management and rank-and-file workers.

8.3.3 Wages and Benefits

Wages are a good basis for comparison in order to understand the nature of the labor market. In the 1998 TVE survey, we asked managers the aggregate amount spent on employee compensation, but it is unclear what employee compensation comprises (i.e., whether it includes all wage and non-wage benefits). Furthermore, among the 158 firms surveyed, only 66 provided answers to this question. Nonetheless, the answers indicate that employee compensation, on average, accounted for 8% of the firms' flow assets (compared to 12% for transportation, 7% for depreciation, and 8% for taxes, among other items). The aggregated amount (total employee compensation) divided by the total number of employees (all categories) yields a crude measure of annual average employee compensation, which amounts to 6,289 RMB/employee/year in 1998. Village-owned TVEs spent the most on employees (7,792 RMB), whereas shareholding and township TVEs spent the least (5,867 RMB and 5,652 RMB, respectively). In general, average wages at SOEs are lower, perhaps partially because of the relatively higher proportion of other benefits.

Although the SOEs in China have historically served both economic and social-welfare functions, the TVEs generally do not bear the burden of social provisions (meals, health, housing, transportation, etc.). Nonetheless, we discovered from our site visits that TVE employees still received some fringe benefits (however minimal), such as medical check-ups and sometimes subsidized lunches at company cafeterias. Some of the coke workers also live in employee quarters on company property, often in close proximity to the cokemaking facilities; however, such living arrangements raise concerns regarding workers' exposure to harmful substances and pollution, and the workers' families are usually not provided housing at these on-site living arrangements.

8.3.4 Health and Safety

Because cokemaking facilities are dangerous environments in which to work, the enterprises and workers should give primary concern to the workers' health and safety. Plant personnel operate the ovens at high temperatures (above 800°C), and the equipment is heavy and cumbersome. Furthermore, the coke ovens, especially those that employ less-advanced technologies, are highly polluting. Coke plants have piles of exposed coal, from which coal dust may be released, and spontaneous combustion can occur (due to the exposure to the sun and trapped heat). During the coking process—and especially at the older coke plants, carcinogenic gases and substances (benzene, toluene, xylene, sulfur dioxide, etc.) are often released into the air, the land, and the water (Polenske and McMichael 2000).

Table 8.2: Employee Safety, 1998, 2000, 2001

Unit: percent

	Safety Instructions	Helmets	Hard Shoes	Eye Goggles	Fire Extinguisher	First-Aid Kit	Other
TVE-1998	79	68	77	60	43	32	9
TVE-2000	91-96	76	66	61	51	41	7
SOE-2001	n.a.	98	78	67	71	49	n.a.

Source: AGS (1998, 2000 and 2001).

Note: n.a. = not available

The most widely used forms of health and safety measures are safety instructions, helmets, hard shoes, and eye-goggles (Table 8.2), and in our survey, managers of SOEs claim to embrace more safety measures than those at TVEs. SOE plant managers require almost all workers to wear helmets and receive safety instructions. However, we noted on our site visits that, in practice, especially in TVEs, these safety measures are not necessarily enforced.

Although the rhetoric and actions concerning worker safety seem to be stronger at the more closely monitored SOEs, we found that in both SOEs and TVEs, there are violations of the aforementioned safety standards as well as pollution regulations. This is not surprising, given the rapid pace of industrialization in China whereby rules and procedures cannot always catch up with growth and expansion, and given that Shanxi Province is a particularly underdeveloped region. In both SOEs and TVEs, the lack of infrastructure and resources for proper monitoring of factories as well as lack of education concerning the dangers of environmental pollution are real problems that will likely be remedied with time and further opening up of the economy to international competition. In addition, both domestic and foreign news agencies are beginning to report abuses, thereby inspiring action. The problems confronting SOEs are likely to differ from those confronting TVEs, but a thorough investigation of the causes of the differences and ways to overcome those problems is beyond the scope of our study.

8.3.5 Labor Implications of Organizational Structure and Technology Changes

When Shanxi Province towns and villages establish TVEs, they open new options for growth and expansion. In this section, we examine how the growth trajectory of TVEs may affect regional socioeconomic conditions.

TVEs are a composite of different types of ownership and are evolving in nature. To study the changing ownership structure of cokemaking TVEs, we examine the sample of 31 common plants (surveyed in 1998 and again in 2000). Cross-tabulating the ownership of the plants in 1998 against their ownership in 2000 yields interesting results (Table 8.3). Most of the plants surveyed underwent a change in ownership between 1998 and 2000, although not according to any noticeable trend. Cross-section data indicate, however, that there is an increase in self-owned and shareholding firms since 1993 (Figure 8.4).

The asset-employment tradeoff involves a balance between investing in capital (e.g., technology, plant facilities, etc.) and investing in human capital (e.g., workers' training)

(Table 8.4). We have two hypotheses that have contradictory implications. First, if labor (human capital) and capital were true substitutes, then higher levels of capital investment would be correlated with low levels of investment in workers' training. This is consistent with a "containment" hypothesis where workers are de-skilled and are stuck with repetitive tasks as management invests in technology. The other hypothesis, however, is that the more profitable and larger a cokemaking plant is, the better it can afford to train workers. Furthermore, advanced equipment that is associated with plant expansion may require more skillful operators than previously, making training a must. Table 8.5 represents a matrix of the tradeoff between the firms' fixed-asset investments and their time investment in labor training. We have only limited data, so that we cannot find an observable pattern of differences across the various investment distributions, but we note that all of the plants that did not provide employee training fall within the low-end of the fixed-assets investment spectrum (100-300 million RMB in 2000).

Table 8.3: Changes in TVE Coke Plant Ownership, Shanxi Province 1998-2000 (31 common plants)

Ownership in 1998 (Number of plants)	Ownership in 2000 (number of coke plants)					
	Township (2)	Village (1)	Joint (1)	Self (12)	Rent/Lease (0)	Shareholding (12)
Township (3)	1			1		1
Village (1)				1		
Joint (2)				2		
Self (12)		1	1	6		4
Rent/Lease (1)						1
Shareholding (9)	1			2		6

Source: AGS (1998 and 2000).

Table 8.4: Trade-offs Between TVE Capital-Expansion and Employment-Level Changes, 2000

Type of Expansion	Increased Employment	Decreased Employment
Expanded Plant		
Number of Plants	33	5
Median Level of Employment in 1995	150	289
Adopted New Equipment		
Number of Plants	24	1
Median Level of Employment in 1995	145	218
Adopted New Technology		
Number of Plants	19	2
Median Level of Employment in 1995	130	179

Source: AGS (2000).

Notes: "Median Level of Employment" refers to the employment level of the plant in the middle of the distribution when plants in the specified category are ranked by the number of employees.

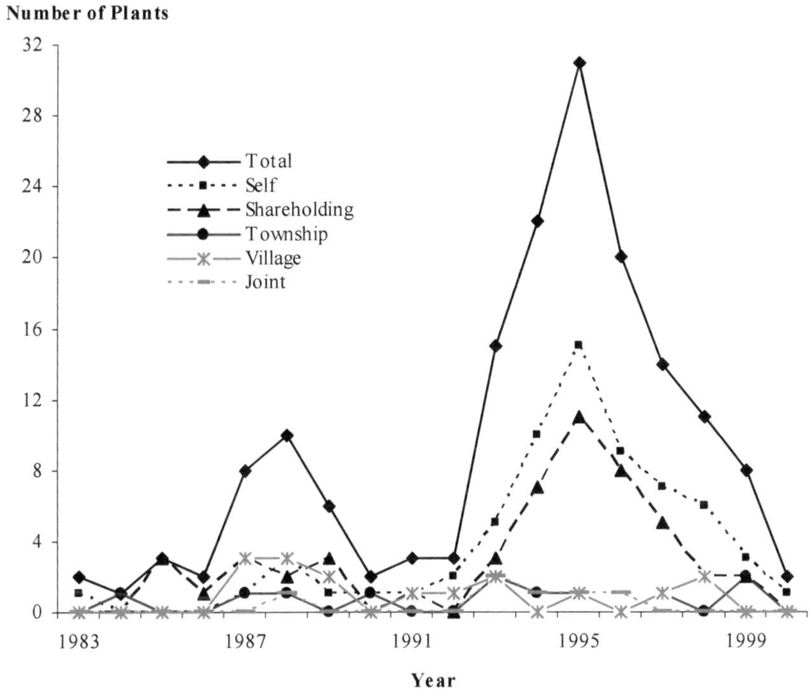

Source: AGS (1998 and 2000).

Figure 8.4: Forms of TVE Ownership by Founding Year of Plant, 1983-2000

Table 8.5: Trade-offs Between TVE Fixed Assets and Employee Training, Shanxi Province, 2000

Fixed Assets Investment in 2000 (Million RMB)	Number of Plants	Training of Production Staff				
		None (%)	1 Day (%)	1 Week (%)	1 Month (%)	Others (%)
0-100	22	9	9	50	18	23
101-200	16	0	19	81	75	25
201-300	10	10	10	30	50	40
301-400	13	0	0	54	38	38
401-500	11	0	0	45	55	9
501-1000	20	5	0	30	60	40
1001-1500	8	0	13	50	38	38
1501-2000	7	0	0	29	43	43
2001-5000	6	0	0	33	50	67
5001 and up	43	7	7	37	21	33
n.a.	30	7	7	37	13	40
All	194	5	6	43	36	34

Source: AGS China Coke Project Survey, 2000.
n.a. = not applicable.
Numbers may not add to 100% due to those not reporting.

Based on the two cases and conversations with managers from TVE plants, we find a trend towards shareholding ownership. Whether these cokemaking enterprises started as farmers' communes or state-owned units, we were told that there is an overwhelming preference for switching towards shareholding in order to generate capital for business expansion. This type of fundraising mechanism and corporate ownership is new in China. On one hand, having diversified their industrial outputs and markets, cokemaking TVEs are less susceptible to market pressures in the coke market. It also reduces risks and increases resilience of the enterprises in the face of rapid technological changes (affecting the demand for coke) and the potential for eventual de-industrialization (affecting the regional economy). Workers enjoy better job security, because they can shift from one industry to another in response to changing product-market conditions. On the other hand, shifting from collective or private, family ownership to shareholding would likely affect the management of the TVEs.

The unique ownership and informal management styles of the TVEs may provide an opportunity for flexible and favorable management-labor interactions, but shareholding ownership is often associated with professional management. Although professional management would take away some of the informality and flexibility that helped TVEs gain a competitive advantage, in the long run, we believe this is a positive trend that would result in better safety standards and wages for workers.

As demand for Chinese coke rapidly increases both domestically and overseas, and as enforcement of environmental regulations increases, coke producers in Shanxi Province will probably be forced to upgrade their equipment and technologies to increase production capacity and to utilize cleaner technologies to meet environmental standards. During his research, Chen Hao (2000) discovered that coke plants that employ new, cleaner coke-oven technologies devote 5.2% of total costs to labor to operate small-machinery ovens compared to 4.7% for JKH-97 (advanced, modified-indigenous) plants, and 3.4% for SJ-96 , i.e., Sanjia-1996 plants.

To expand this line of research one step further, we look at the relationship between average labor compensation (per capita) and technology used at the plants among the 158 plants surveyed in 1998. A plant using the PX (Pingxiang) oven fell in the lower end of the spectrum, spending 1,117 RMB per employee per year in compensation. Employees at nonrecovery (SJ-96, i.e., Sanjia-96)) plants had a below-average compensation of 5,832 RMB/year, compared to the high end of 10,526 RMB/employee/year at 2H-II ovens, which are small-machinery slot ovens (Figure 8.5). Overall, it appears that except for firms using indigenous technologies (namely, PX ovens) that pay employees significantly less, the relationship between technology choice and employee compensation is unclear based on the data. Unfortunately, due to the small sample sizes for each technology, there are high degrees of variance attached to these figures. Even so, we have shown that the product diversification and ownership trends have labor-market and management implications.

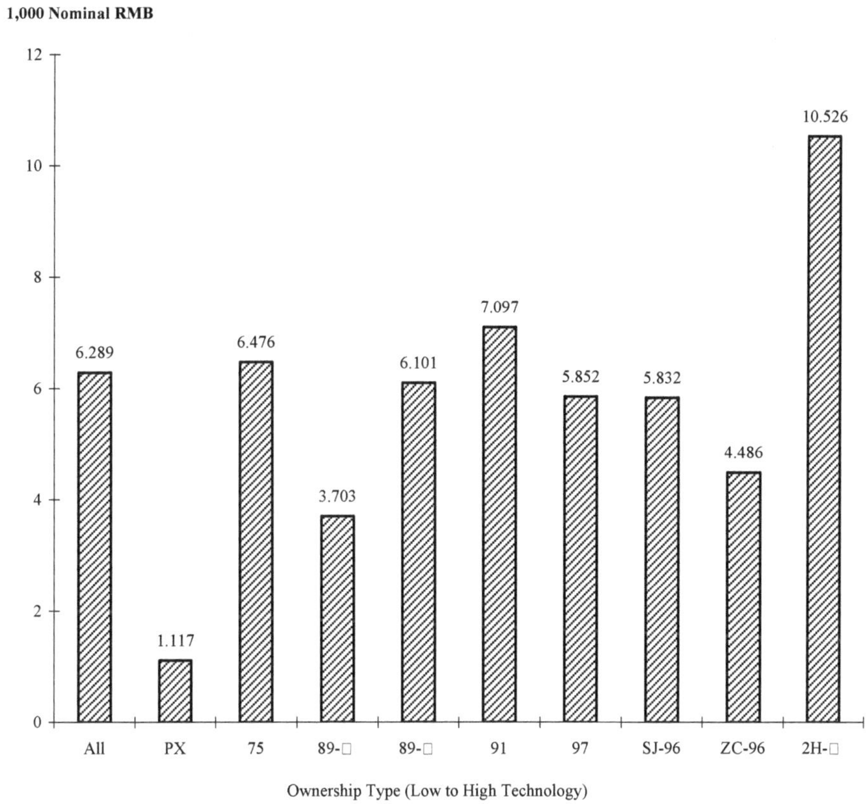

Source: AGS China Coke Survey, 1998.

Figure 8.5: TVE Coke-Oven Technologies and Employee Compensation, Shanxi Province, 1998

8.4 Conclusion

In the past two decades, cokemaking has generated sizeable amounts of income and industrial employment in Shanxi Province. As China becomes an increasingly influential player in the global coke market, as Chinese domestic enterprises undergo rapid institutional reforms, and as new technologies and environmental policies are introduced in Shanxi Province, the cokemaking sector reacts to these challenges and adapts accordingly. The sector's responses, in turn, impact the livelihoods and health of the region's residents and workers, and they can affect the sustainability and resilience of the regional economy.

We have shown that because TVEs tend to be relatively very labor intensive, their recent proliferation has had significant impacts on rural employment in Shanxi Province. As long as TVEs continue to have a competitive advantage over SOEs relying on cheap rural labor, TVEs will be more effective in generating rural industrial employment than SOEs. But this situation may not last very long–as TVEs generate more revenue through investment in capital improvements, they will rely less and less on unskilled rural labor.

Although TVEs have been successful in raising income and employment in the countryside in Shanxi Province, analysts could examine the quality of the work life at these enterprises and their long-term socioeconomic sustainability. In addition to providing first-time industrial employment opportunities, we wanted to see whether TVEs also provided opportunities for professional advancement in terms of education and training. Although we found that education levels of workers at TVEs were generally higher than the provincial population average, we also found that the amount of on-the-job training has been decreasing in recent years (e.g., only 44% of surveyed TVE production employees in 2000 received one week or more of training, compared to 78% in 1998). Meanwhile, training opportunities at SOEs have been on the rise. We also found that while wages at TVEs (as a percentage of revenue) are generally higher than at SOEs, benefits such as medical care and training are less substantial. Further, while workers' health and safety are concerns at both SOEs and TVEs, SOEs seem better at enforcing pollution and safety standards and providing workers with medical care.

Finally, we examined how changes of ownership structures and technology changes are impacting the welfare of workers and concluded that trends towards product diversification and professional management can improve job security for workers, increase the number of avenues for advancement, and improve safety conditions. In some ways, these changes are not surprising in that they mirror industrial trends that took place in developed countries many decades ago. In other ways, they are surprising, because many improvements in worker conditions in industrialized countries have happened through the collective organization of workers, which is an option that is not available to industrial workers in China. Chinese unions are government-sponsored. There are few, if any, legally recognized unions that are organized and run entirely by workers.

We hope that through this work, we have been able to provide a human face to the cokemaking industrial trends and technological changes that have been discussed in previous chapters of this book. Although it is important to understand how such rapid and sweeping changes impact GDP and energy intensity, we believe it is equally important to see how these trends directly affect the lives of the very people who are helping to realize these changes.

References

Cable, Vincent. 1996. "The Outlook for Labour-Intensive Manufacturing in China," in *China in the 21st Century*, published by the Organization for Economic Co-operation and Development (OECD): 37- 52.

Chan, Anita. 2000. "Trade Unions and Workplace Relation." Chapter 3 in Changing Workplace Relations in the Chinese Economy, edited by Malcolm Warner. London, UK: MacMillan Press Ltd.: 34- 56.

Chan, Nelson. 1999. "Land-Use Rights in Mainland China: Problems and Recommendations for Improvement." *Journal of Real Estate Literature*. 7(1) : 53-63.

Chen, Hao. 2000. *Technological Evaluation and Policy Analysis for Cokemaking: A Case Study of Cokemaking Plants in Shanxi Province, China.* Masters thesis submitted to the Massachusetts Institute of Technology, May 2000.

ChinaOnline. 2000. *Township and Village Enterprises*. May 31, 2000. Available online at: http://www.chinaonline.com/refer/ministry_profiles/moa-tve.asp

Chinese Statistical Yearbook. Various Volumes. Beijing, China: State Statistical Bureau.

China Steel Yearbook. 2000. Beijing, China. The Ministry of Metals.

China Township and Village Enterprise Yearbook, 1998-2002. China Agricultural Press.

Fang, Jinghua. 2002a. Internview by Hoi-Yan Erica Chan at the Alliance for Global Sustainability Annual Conference in San Jose, Costa Rica, on March 20, 2002.

Fang, Jinghua. 2002b. E-mail correspondence with Hoi-Yan Erica Chan, dated April 20, 2002.

Mahdavi, Ali. 2001. "Sources of Economic Efficiency for China's Township and Village Enterprises." Paper presented at the China-Coke Workshop, Alliance for Global Sustainability Conference in Lausanne, Switzerland. January 14, 2001.

New China News Agency (*Xinhua*). 2002. "China ABCs". Available online at: http://202.84.17.11/english/china_abc/taiyuan.htm

Polenske, Karen R., and Francis McMichael. 2002. "A Chinese Cokemaking Process-Flow Model for Energy and Environmental Analyses." *Energy Policy*. 30(10): 865-883.

Shanxi Statistical Yearbook, 2001. Beijing, China: China Statistics Press.

United Nations Economic and Social Commission for Asia and the Pacific (UN-ESCAP). 2002. *Population Programme databases: Shanxi*. Available online at: http://www.unescap.org/pop/database/chinadata/shanxi.htm

---

[1] Currently, Consultant, Business Strategy and Policy Group, Steer Davies Gleave, London, England; at time of writing, Multiregional Planning (MRP) Research staff, MIT, Cambridge, MA, USA.

[2] Multiregional Planning (MRP) Research staff, MIT, Cambridge, MA, USA.

CHAPTER 9

CONCLUSION

Karen R. POLENSKE[1]

9.0 Summary

In the preceding chapters, members of the China cokemaking team describe the evolution of the coke sector in the People's Republic of China since the late 1990s, with emphasis on cokemaking in Shanxi Province. In this chapter, first, I summarize the seven-year study by focusing on the key issues confronting the cokemaking sector at this fascinating juncture in China's development. Then, I describe some of the major achievements of our China cokemaking team as we have been fortunate to observe and record many of the dramatic changes occurring in the coke sector in Shanxi Province during part of its transformation from being one of the most antiquated coke-manufacturing sectors in the world to becoming one of the most modern ones.

Finally, I set forth the potential applications for our analytical techniques to other sectors, regions, and countries. I also discuss the critical need for extending the present study. I propose three possible extensions: (1) Examination of the health of the workers and people in the communities near to coke plants; (2) Comparison of cokemaking in Shanxi Province with that in other parts of China and an even more important comparison of cokemaking technologies in China with the technologies used in other countries, such as Brazil, India, Japan, and the United States; and (3) By changing the focus somewhat, examination of the need to recycle the land upon which the coke plants that are being closed are situated in order to turn that land and those neighborhoods into a sustainable economic asset for the province.

9.1 Key Cokemaking Issues

Throughout this book, we document the dramatic rise of coke production by Township and Village Enterprises (TVEs) in China as coke production is shifting from state-owned enterprises (SOEs) to production by TVEs and private and stockholder

Karen R. Polenske (ed.), The Technology-Energy-Environmental-Health (TEEH) Chain in China: A Case Study of Cokemaking, 149–158.

companies. In fact, the names SOE and TVE are being phased out. At the beginning of the 1990s, SOE coke production represented close to 80% of total coke production in Shanxi Province and TVE coke production 20%, but by the early 2000s just the reverse was the case, as TVE coke production boomed. This dramatic shift accompanied by changes in technology, have had far-reaching impacts on energy use and resource consumption, the environment, and human health. We characterize these impacts as being part of a more general chain: **t**echnology, **e**nergy, **e**nvironment, and **h**ealth (TEEH), which I review in more detail in the following section.

9.1.1. Technology

In Shanxi Province, coke-plant managers have been replacing indigenous, modified indigenous, and small machinery ovens with modern, large-machinery, as well as clean-coke ovens (Chapters 2 and 3). These technology changes have reduced the energy intensity in cokemaking (Chapter 4) and have brought about three significant changes in the sector. First, technology changes have reduced the need for low-skilled workers; second, such changes have affected the amount of capital required to produce coke; and third, they have expanded the potential markets served by the Shanxi coke producers.

Employment Effects. Compared with the coke ovens used previous to 1995, the newer (2001-2004) ovens possess considerably larger output capacity, are more energy efficient (tonnes of coal per tonne of coke), environmentally efficient (tonnes of pollutant per tonne of coke), and labor efficient (numbers of workers per tonne of coke). We have documented the increase in workers' productivity. As production shifts from SOEs to TVEs, we might expect more coke workers to come from the rural areas, where most TVE plants are located. The TVEs tend to use simpler production technologies than SOEs, but the peasants in the rural areas are usually farmers with limited education, thus not trained to do the highly skilled work required in modern coke plants.

The shift in cokemaking technology is creating a need for more highly skilled workers than in the past, implying that many TVEs will need either to train their current employees or hire workers who are not from the surrounding countryside (Chapters 5 and 8). Although older-technology TVEs provide much-needed low-skill industrial jobs in rural areas, they are also highly polluting and harmful to the environment and human health. Thus, local officials are confronted with deciding whether to encourage the TVEs to adopt new technologies that would require a well-educated work force or to shut the plants down entirely. Unless training workshops are established especially in rural areas, either solution may create unemployment for the former coke workers who are relatively unskilled and many of whom still live in rural towns and villages.

In 1998, Shanxi Province had 1,872 coke plants; by 2003, this number was reduced to 589 plants, implying that many managers, willingly or unwillingly, closed their plants. This meant that a significant number of workers, i.e., mainly peasants, have lost jobs and income, and we have heard scattered reports of peasant unrest in some towns and villages, perhaps as a result of these job losses. Thus, the decisions for the cokemaking sector will affect what employment opportunities remain for peasants and what types of skill-enhancing means plant managers establish to help workers train for new, skilled work in the plants.

Capital-Requirement Effects.
We have shown that although more than 1200 coke plants in Shanxi Province have been closed between 1998 and 2003, coke output has remained relatively constant during those years. In 2003 and 2004, coke production even increased as some managers either built new high-capacity plants or expanded the production capacity at old plants. Either solution has augmented the need by coke-plant managers for capital to invest in new technologies. Local plant managers in rural areas usually do not have easy access to loans and other funds from state financial institutions either to purchase new, usually low-pollution, ovens or to upgrade old ovens to meet local pollution regulations. We have not conducted a systematic study of the capital markets in the rural areas, but coke managers and others we interviewed during the past seven years have stressed the difficulties in getting easy access to funds for necessary upgrades and expansion. Anecdotal evidence is that capital comes from local sources (families and friends). Obviously, because such an expansion is being undertaken by some managers, they have found ingenious ways to obtain the funds.

Coke-Market Effects.
Changes in technologies also affect the potential markets served by the Shanxi coke producers. Through our surveys, we are not only able to know where the coke is marketed (within or outside the region) but also to what intermediate or final consumer it is marketed, as well as the distance and location of the coal mines from which the plants obtain the coal. Through the use of a geographic information systems (GIS) planning-support system, our team members have traced the flow along the cokemaking supply chain from the coal mine, to the coke plant, and to the final end user (Chapter 6). We have learned that the nature of the end users has been changing considerably, as foreign markets rapidly expand and domestic demand increases. Because rail-transport capacity is often not available, trucks are being increasingly used even for long-distance transport to the ports, especially to Tianjin and Qinhuangdao. Given these significant trends, we expect our understanding of shifts in the coke market to become increasingly relevant in the coming years.

As China's own development needs increase, more and more of the coke is needed by steel producers in different parts of the country, We see a need for a major study concerning what local (Shanxi Province), regional (Northeast China), and national (China) markets as well as international markets can be served by the current and future producers in Shanxi Province, the largest coke-production region in China. Also, we need information on implications different coke markets will have on the transportation needs for the country as well as on the energy requirements and pollution generation created by alternative transportation solutions, some of which I discuss next.

9.1.2. Energy Consumption

Coal consumption per tonne of coke produced has declined from 1978-2003 in Shanxi Province, China (Chapter 1-3). Although the primitive indigenous ovens used about 9 tonnes of coal per tonne of coke, the latest technology ovens use only about 1.5-3.0 tonnes of coal per tonne of coke. Even so, total use of metallurgical (coking) coal has increased dramatically as the total output of coke in Shanxi Province increased from 16 million tonnes in 1990 to almost 70 million tonnes in 2003. Coking coal is in short supply worldwide, and Shanxi Province has some of the largest supplies of this valuable commodity. Coal is a fossil fuel, which is not renewable. Will there be sufficient metallurgical coal for the rapid industrial expansion China is experiencing if such a rapid growth is maintained for the next 20 years? If not, what alternatives can producers depend upon for production of iron and steel and related products and how will this affect the production of construction materials, automobiles, other consumer durables, and other products that currently need steel, thus coke? These questions have arisen as we worked on this study.

We designed the Shanxi Province Geographic Planning Support System (SPGPSS) and applied it to the analysis of transportation and location choices faced by the coke managers and local government officials in Shanxi Province, China (Chapter 6). Decision makers in Shanxi Province can use this model, as we did, to examine alternatives to minimize production and transportation costs, total energy consumption, and total pollution emissions from producing coke.: Analysts can also adapt the model for use with other sectors, other regions, and/or in other countries.

9.1.3. Environment

Pollution per tonne of coke produced, on the one hand, has increased because cokemaking TVEs produce more pollution than cokemaking SOEs. On the other hand, it has decreased due to the closing by local authorities of the most polluting indigenous and modified-indigenous plants. We have not been able to measure precisely the net effect, because we do not have an exact count of each type of coke oven in Shanxi Province, nor of the amount of output produced and pollution created. Instead, for our work on the environment, we have concentrated on measuring the ultra-fine particulates initially at the plant during cokemaking and have extended our analysis to include measurements from the transport of the coke as well as in the coke-worker homes. (Chapter 7).

In Shanxi Province, we used mobile sensors of polycyclic aromatic hydrocarbons (PAH) generated by coke ovens, diesel trucks, and other sources. A member of our cokemaking team, Professor Hans Siegmann, designed and constructed the first two types of mobile PAH sensors that we used for this research. Using those sensors, which are now commercially available, as well as an already-available commercial large-particulate sensor, we conducted a series of tests over the seven years of the project to measure PM_{10} and $PM_{2.5}$ emissions not only from different types of coke ovens, but also from the diesel trucks used to haul the coal to the coke plant and the coke from the plant to the customer,. We can pinpoint the precise times (e.g., during a coke push) and locations coke workers are at most risk of breathing in harmful substances, and can say, interestingly, the relative degree of risk at these different times and locations. We also used the sensors to measure the level of particulate pollution in coke workers' homes.

9.1.4 Health

From our pollution studies, reported in Chapter 7, we know that significant amounts of particulate pollution occur not only from the coke ovens and surrounding activities at the coke plant, but also from the transport of the coal and coke by truck. Based on studies in other countries of the health effects of ultrafine particulates from combustion sources, we know that pollution from both sources may be causing serious health problems not only for the workers, but also for the residents in the towns and villages near to the coke plants. In order to complete our study of the TEEH chain, we would need to conduct a health-impact study in Shanxi Province, something that has been beyond the scope of the funding we have obtained to date.

9.1.5 Final Remarks on the TEEH Chain

In earlier chapters, we documented the types of new technologies being adopted both for the large-machinery, slot-oven batteries and the different types of so-called "clean" coke ovens. We believe that there is a need for a continuation of the type of surveys we have taken in order to maintain the historical record of the type of ovens in operation at each plant, the amount of energy and other inputs used, the location of the coal suppliers and coke markets, and the changes in these and other factors over time. With such information, analysts can trace not only the technologies that are being used to produce coke, but also the energy use, transportation needs, pollution changes, and market shifts occurring in this rapidly changing sector.

9.2 Major Achievements of the China Cokemaking Team

We are working at the forefront of the environmental field in terms both of adapting current and developing new economic, transportation, and pollution-monitoring tools of analysis. During the past seven years, members of our multidisciplinary team, which comprises chemical engineers, economic planners, and physicists, have been developing the tools and methods necessary to understand the complexities of the cokemaking sector through work in their own universities and from the regular (once-or-twice-a-year field trips to China. We have used the primary research funding from the Alliance for Global Sustainability (AGS) to leverage several other important funding sources, including the U.S. National Science Foundation. As of this year (2005), our findings are being reviewed and discussed at all levels of government in China.

We have developed a set of unique methods of combining micro energy, employment, technology, and transportation data from our plant surveys with macro data from published statistics. The following five distinctive tools of analysis provide us with a powerful means of communicating critical information to planning officials at the plant, local, regional, and national officials, and to analysts in chemical engineering, economics, geography, planning, and transportation disciplines.

9.2.1 Plant and Local-Official Surveys

We use time-series surveys to track the critical variables on energy use and pollution generation year-by-year at the plant level. Most other research groups in China and other emerging countries do a survey for only one or two years. Since 1998, we have conducted seven surveys, which provide us with the unique ability to analyze effects of changing environmental policies on employment, energy use, pollution, and transportation. This is important information, some of which we present in this book, to help track the dramatic transformations occurring within the coke sector in Shanxi Province.

9.2.2 Case Studies of Individual Plants

We use case studies to verify the information from the surveys and to obtain in-depth technology information that helps us understand how the adoption of alternative technologies is affecting the economic base of the region. We combine plant case studies, plant-manager survey data, and aggregate analyses for Shanxi Province and China to help us understand how technological changes are affecting energy intensity.

9.2.3 GIS-based Planning Support System (GPSS)

We use the GPSS to study how different technology and plant-location policies affect total energy use, costs, and pollution. We have developed the system so that it could be used by planning officials in Shanxi Province or by other analysts to examine how alternative locations of sources of coal, as well as plants and transport facilities may affect the energy used and pollution generated in making coke. Analysts can easily adapt the system to other regions and sectors in China or in other countries.

9.2.4 Extended National and Provincial Input-Output Tables

We have extended the national and provincial input-output tables in two ways: first to examine the detailed labor, capital, and land-use requirements of alternative technologies, and second to examine the differences in technologies between state-owned enterprises (SOEs) and township and village enterprises (TVEs), with an interest in understanding technology differences in large-scale and small-scale plants and in plants with different types of ownership (state versus nonstate). This is one of the first times that analysts have examined technologies of different sized plants and other economic activities in a systematic way within a region as well as to make comparisons with the technologies in China as a whole.

9.2.5 Particulate Air-Pollution Sensors

One member of our team, Professor Hans C. Siegmann, designed and developed two of the three mobile air-pollution sensors we use to determine particulate-pollution concentrations. We also employ a third already commercially available sensor. to examine large-sized particulate pollution. In 1998, we conducted the first testing of these sensors in Shanxi Province in the coke plants and in the homes of the coke workers, and we have continued to use them throughout our study. During the first six years of our research, we had some difficulty getting access to plants to do appropriate testing, but in 2004, we were actually invited to conduct tests by plant managers who had installed clean coke ovens and wanted to know how well the ovens performed. The sensors we have used in our work have also been used for the Alliance for Global Sustainability Mexico City Project (Molina and Molina, 2002), and we have loaned one of our sensors for a demonstration in a new project on community-pollution efforts in the United States.

Thus, these five distinctive tools of analysis are being used not only for our own analyses, but they are increasingly being adopted by other analysts for use in related studies.

9.3 Potential Future Studies

In the future, we have many ways to apply our analytical techniques to other sectors, regions, and/or countries and to extend our own studies to examine the health of the workers and people in the communities near to coke plants. In addition, we have an intriguing possibility of comparing cokemaking in Shanxi Province to that in other parts of China as well as to that in other countries, such as Brazil, India, Japan, and the United States.

9.3.1 Health Studies

To complete the TEEH chain work we began in 1997, we would like to examine the health effects of cokemaking on coke workers and their families. A comprehensive health study would require (1) five or more years of tracking the coke workers and their families, (2) cooperation from a local medical university, and (3) more pollution sensors and personnel to conduct the tests. We have designed a study that would involve working with medical personnel at the Shanxi Medical University, who would conduct the actual medical tests in Shanxi Province, and environmental and occupational epidemiologists at the Harvard School of Public Health (HSPH), who are very experienced in conducting health studies in China.

Should funds become available, we would work with these analysts to conduct a health survey to estimate the exposure-risk in the coke working population and nearby communities. The survey would have four specific objectives: (a) to determine the prevalence and incidence of acute and chronic symptoms and diseases, focusing on the respiratory system, in the coke-plant workers; (b) to examine the clinical and sub-clinical respiratory effects of air pollution due to coking emissions in local communities; (c) to carry out a prospective follow-up study in order to observe the chronic respiratory effects of a persistent or long-term exposure; and (d) to assess potential exposure-response relationships. During the survey, we would take note of whether subjects are using protective gear, and the relative degree to which plants comply with provincial emissions standards. We hypothesize that (1) substantial respiratory problems can be caused by long-term heavy exposure to cokemaking emissions; (2) these effects are felt both by workers and by people in nearby communities; and that (3) the severity of impact can be mitigated by the use of protective gear and better adherence to emissions standards. This work would provide the data necessary to promote action in improving worker safety, worker medical care and health insurance standards, and enforcement of pollution standards. To make a complete environmental health study of cokemaking would be to include a study of the interrelations between water pollution and health as well as the interrelations between land pollution and health. Such a study would complete the TEEH chain with which we began our work in 1997.

9.3.2 Comparative Studies

We believe that the following studies are critical in order to push the frontier of the state-of-the-art in this field:

At the current time (2005), although China is a major global producer of coke, we think it is important to examine coke production in other countries to see the extent to which any of the lessons China is learning are applicable elsewhere. We have tentatively selected the following countries for such a comparative study:

Brazil is an interesting country to study, because they use charcoal rather than coke for about 15% of the iron made in the iron blast furnace. Charcoal was used prior to coke in early ironmaking in China, Sweden, England and elsewhere. What issues does the coke sector in Brazil confront when there is another competitor to coke? What effect does the use of charcoal have on the metallurgical coal market in Brazil?

India is a county that currently has about the same population as China, and that is expected within the next 20 years to surpass China in population. Also, India has a long history of coke and iron and steel production. In addition, some Chinese coke technology is being transferred to India, so that it is appropriate to see what the coal, coke, and steel sectors in India are using in terms of technology and to determine what opportunities exist for such a South-South technology transfer to occur.

Japan, is a major foreign consumer of Chinese coke, yet also uses modern large-machinery slot coke ovens to make their own coke. Many plants uses dry quenching to cool the coke, which saves a considerable amount of water. We have learned that a few coke plants in Shanxi Province are already using or are considering the use of dry quenching. Because water is in very short supply in Shanxi Province, we are interested in determining how transferable this and other Japanese technologies are to Shanxi Province.

United States, clean coke ovens similar to ones now used in China were first introduced in the United States. Are there possibilities for further expansion of this technology in the United States? To what extent may the burning of all the chemical by-products lead to future shortages of these products in the United States or in China?

Both **India** and China possess very significant coal reserves, as well as populations significantly large enough to exploit them. Pollution, greenhouse gas emissions, and energy security are among many concerns that all these countries have in common. We are interested in comparing how the countries are attempting to tackle these issues–issues that could have far-reaching impacts on world energy markets, global warming, and air pollution.

9.3.3 Recycling Land

Within a five-year period, from 1998 to 2003, at least 1,283 coke plants were shut down in Shanxi Province alone. Since cokemaking in China typically pollutes surrounding land and water, these plant closures represent a significant proliferation of land that could be recycled. In the United States and some other countries this land is called a "brownfield," and is defined as land parcels that are abandoned or have a long vacancy, but in China, there is not even a word for "brownfields." Most developed countries have well-established laws and policies regarding brownfields (e.g., remediation standards, specific liability laws, and clean-up grant programs), Without remediation standards, developers are free to build on highly contaminated soils. In addition, without appropriate liability laws or clean-up programs, there is little incentive on the part of plants to prevent ground pollution or to cleanup the pollution once it occurs. As urban areas in China expand rapidly and property markets develop, land values are increasing dramatically. Thus, with the mass closure of polluting factories in China, brownfield redevelopment will become an increasingly pertinent issue in years to come. We could draw upon our vast knowledge of plant pollution, ownership structures, and environmental regulations to examine how the brownfields' question might be most appropriately analyzed in the Chinese context, especially in cokemaking.

These are the studies that we think have the most importance at the present time related to the coke sector.

9.4 Conclusion

In our research, we have drawn upon a wide-range of analytical techniques, such as pollution monitoring, plant surveys, economic analyses, and simulation modeling of new manufacturing and transportation technologies. To bridge these different methods of inquiry, which are conventionally treated separately, is a risky undertaking not generally supported by granting agencies. We were fortunate that the Alliance for Global Sustainability as well as the U.S. National Science Foundation and the universities or academic units with which our project members are affiliated have generously contributed funds to this research.

As part of the AGS research, we have gradually developed a large network of interested stakeholders in China. This network includes national officials in Agenda 21, National Environmental Planning Agency (NEPA), and the Ministry of Agriculture; local environmental and planning officials; academics at the Taiyuan University of Technology and the Chinese Academy of Sciences in Beijing; and cokemaking plant managers and other plant officials.

Our integrated, cross-disciplinary approach has enabled us to provide stimulating insights into cokemaking in China that are central to determining how human (economic) activities can achieve a sustainable economy. This has been an exciting and rewarding research project, and we hope that we have been able to convey its importance to the reader.

References

Molina, Luisa T., and Mario J. Molina, eds. *Air Quality in the Mexico Megacity: An Integrated Assessment.* Dordrecht, The Netherlands: Kluwer Academic Publishers.

---

[1] Professor of Regional Political Economy and Planning. Head, China Cokemaking Team, Department of Urban Studies and Planning, Massachusetts Institute of Technology, USA.

APPENDIX A

PHOTOS AND DESCRIPTIONS

COKEMAKING PHOTOS, SHANXI PROVINCE, 1998-2004

Karen R. POLENSKE

A.0 Photo History of Cokemaking in Shanxi Province, 1998-2004

Since we took our first field trip in July 1998, our cokemaking team has taken many photos, a few of which I include in Appendix A. I note those not taken by me. Little did I realize during that first field trip that I would be recording the beginning of a rapid transformation in the cokemaking technologies in Shanxi Province. I started to photograph at the coke plants in Shanxi Province in 1998, and I have used the more than 2,000 photos I have taken in many of my powerpoint presentations, as well as in the posters that we display at the poster session for the Annual AGS conference, and to show the research assistants (RAs) some of what to expect when new RAs prepare for their first field trip. In the United States, I have shown them at the different coke plants I have visited to illustrate some of the transformation that is occurring in China. I try always to ask for permission before taking any photo. Only a few times have I been told not to take a photo. In the photos, note that the picture often seems hazy. The photo is not bad, but it is the polluted air that is causing the photo to look hazy. In this Appendix, I share with you, select photos to help tell the remarkable story of the TEEH chain in Shanxi Province. Several of the ovens, batteries, and other items I have photographed no longer exist.

I call the maps "Photos," just for ease in numbering. In Photo 1 (created by Li Xin in June 2005), I provide the map of China and show where Shanxi Province and Beijing are located. In Photos 2 and 3, I show two road and rail maps of Shanxi Province (Chapter 6), identifying the coal mines, coke plants, and the major highways and rail lines. Photo 2 shows coal shipments, and Photo 3 shows coke shipments in the region, both for the year 2000.

Karen R. Polenske (ed.), The Technology-Energy-Environmental-Health (TEEH) Chain in China: A Case Study of Cokemaking, 159–178.

A.1 Coke-Oven Technologies

In Photo 4, I show the line of coke chimneys that we saw on our first 150-Kilometer trip from Taiyuan to Hongtong. In this first section of photos, I document different coke-oven technologies from the hole-in-the-ground indigenous ovens to the modern slot ovens and clean coke ovens. In Photo 5, I show an indigenous coke-oven site that we visited south of Jiexiu on our first field trip; and, in Photo 6, I show five men who are manually pushing a cover onto a modified indigenous oven. This oven was torn down at the Antai Coke plant in Jiexiu the next year and replaced by a hotel. In Photo 7, I show batteries of modified indigenous ovens, which probably are all closed. We were traveling on the highway, so that I could not ascertain whether they were operating. In Photo 9, a man is sweeping the area in front of a JKH modified indigenous oven at Sansheng, and in Photo 10, I show the coke oven doors that were to be installed in a new machinery (slot-oven) battery at Qingxu. One battery of slot ovens and the coal-loading tower are in the background. I show the sequence of cokemaking operations from the coke push into the quenching car (Photo 11), to the quenching tower (Photo 13), and to the coke-cooling wharf (Photo 12). The black smoke from the coke push contrasts with the white steam from the cooling tower. I note that the following year, some plants, like Antai, had installed covers on at least some of their quenching cars, as continuous efforts are made by them and other coke plants to reduce pollution. I do not show this in the photos.

In the final set of photos on coke-oven technologies, I show different versions of "clean" coke ovens, originally referred to as nonrecovery ovens. We first saw the SJ-96 (SanJia-1996) ovens, Photos 14 and 15, in 1998 at the Sanjia plant in Jiexiu. The workers sealed the oven doors with mud. Note the blue-painted chimneys in the background. The Environmental Protection Bureau would allow a chimney to be painted blue or green only if the plant met the then increasingly stringent pollution standards. In Photo 16, I show the heat-recovery version of this type of oven, which had been installed at Sanjia by July 2002. In 1998, I heard about the Houma coke nonrecovery ovens (Photo 17), but I did not see them until 2004, by which time they were not operating properly and were being replaced by newer ovens (not shown). Both the Gangyuan coke plant (Photo 18) and the Qingxu Yingxian coke plant (Photos 19 and 20) recently (2003-2004) installed "clean" coke ovens. At the Gangyuan battery in January 2004, the manager asked us to do an intensive particulate-pollution test later in the year. This is the first time a plant manager had asked us to do the tests instead of our asking if we could conduct tests.

The product of cokemaking is coke. I show the difference in size of the metallurgical coke and stacked foundry coke in Photo 21, the workers stacking the foundry coke in Photo 22, and one of many uses of some waste products—in this case to make sidewalk tiles (Photo 23).

A.2 Energy: Coal, its mining, transport, and processing

In Shanxi Province, plants obtain coal from nearby coal mines. We visited only one such mine, which was a pit mine. In 2002, we visited one of the largest open-surface mines in China at Fushun, Liaoning Province. I include Photo 24 to show the extensive rail system they use within the mine. In Shanxi Province, most of the coal is mined from underground mines (Photo 25) and trucked to the plant, where the trucks are weighed (Photo 26) prior to being unloaded. At the Houma Coke Plant, they were using large equipment to unload the coal from the train or truck onto a pile of coal for storing for use in the ovens (Photo 27). At Qingxu Yingxian, the coal is top loaded into the clean coke ovens (Photo 28). Before being loaded into the coke oven, the coal is washed to remove dirt and other impurities (Photo 29). The washing also removes some of the ash from the coal. After the coal is washed, workers test the coal for quality (Photo 30).

For materials, such as brick, for the ovens, workers use animals to pull wagons (Photo 31, taken by Francis C. McMichael). For the movement of the coke on the plant site, they may use small tractors (Photo 32) or local peasants build up the sides of their trailers in order to use them for transporting coke to the local train station. Here, they wait in line to weigh their loads (Photo 33). I show some of the coke cars from a train that passes by a coke plant as it heads towards Taiyuan or other points East (Photo 34). Plants also use large diesel trucks. This truck is transporting foundry coke from a plant to points unknown (Photo 35). Most workers at a plant use motorcycles or bicycles to travel from their nearby villages or towns (Photos 36 and 37).

The year 2000 was one of the first years in which plants used 20-tonne containers for shipping coke to the ports. They load the container manually at the plant and take it to the train station, or they load the container directly onto a truck bed. One man takes one day to load such a container (Photo 38). Two years later, several men were loading one container (Photo 39). The plant used this control room to track the coal and coke. The control-room manager had three red telephones, but we did not determine for what important calls they were used (Photo 40).

A.3 Environmental Pollution Measures and Health

After the slot coke-oven is filled with coal, the men use a sealant to prevent emissions from escaping the oven during the coking process (Photo 41). Different signs are used to warn workers (and visitors) of the dangers at the plant, stressing the need to wear a gas mask (Photo 42) and the dangers of Black Lung Disease (Photo 43). Photo 44 shows the technical drawings of the three sensors members of our team used to detect particulate pollution. Many plants surround coke ovens with trees and grass (Photo 45). Diesel trucks carrying coal must be covered if they use expressways (Photo 47). In Photo 46, I show Dr. Qian Zhiqiang, one of our China coke team members, examining the three types of particulate sensors we used, which are shown in more detail in Photo 48.

Although we have not had the funding to start the health phase of our work yet, we expect that the health of many workers and their families are improving, especially since 1998 as many of the most polluting plants are being closed. For those still open, many coke plants have considerable amounts of green grass, trees, and shrubs, some of which are directly by the coke ovens (Photos 49 and 50). Some plants even have vast beds of flowers (not shown). Indirectly, such flower beds may help the mental attitude of workers as they arrive at or leave work. Whereas the quenching water used to flow directly into the ground at the indigenous and modified indigenous coke ovens, some coke plants have recycling units for the water, which is then reused (Photo 51). These are just a few of the many examples we have seen of the efforts coke managers are making to provide a healthy and safe place in which their employees can work and live.

Even so, the pollution problem is severe. In addition to closing the most polluting plants, the managers are installing newer coke-oven technologies that reduce the pollution emitted from the ovens and quenching cars, but I remain troubled by the lack of attention to the pollution caused by the diesel trucks. Although Shanxi Province authorities have closed two-thirds of the most polluting coke ovens that existed in 1998, the air in the villages and towns where coke production occurs is still hazy most days, and using our three mobile battery-operated particulate-pollution sensors, we continue to detect high levels of particulate pollution in the ambient air, especially along the highways. I am confident that the officials will soon make concerted efforts to solve the freight-transportation problem, so that in the future, my pictures can show blue rather than hazy sky, and, even more important, so that the workers and people in the towns and villages in Shanxi Province can be healthy.

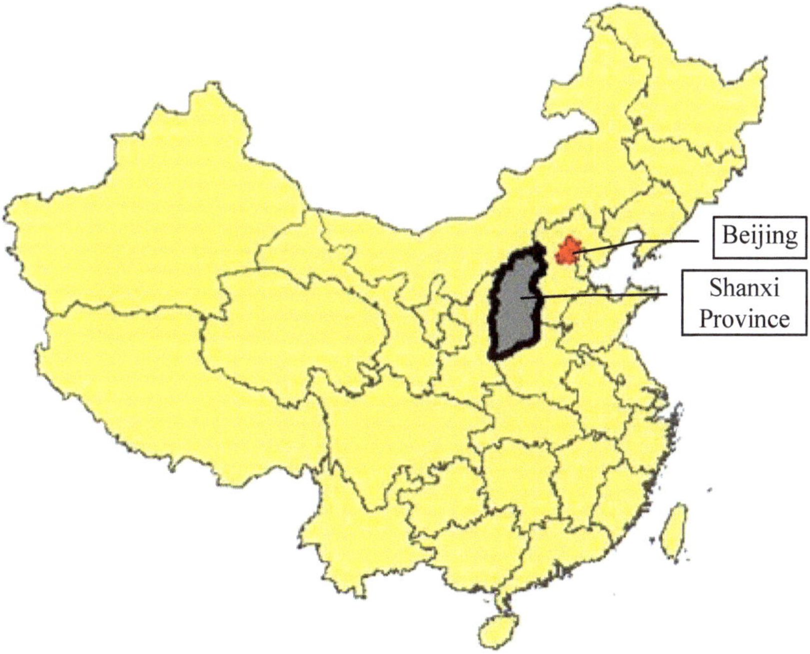

Photo 1: Map of China with Shanxi Province Highlighted

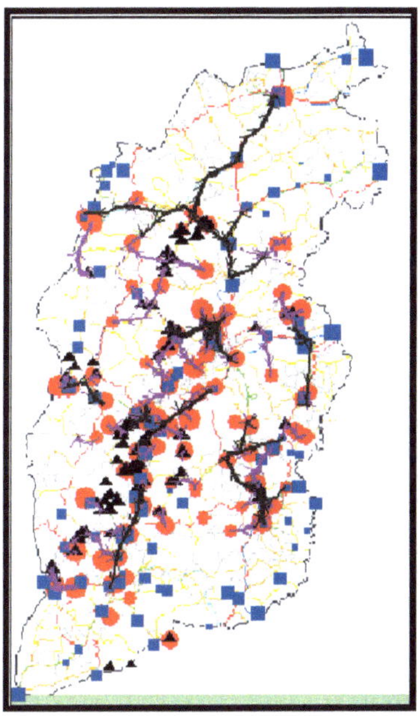

Photo 2: Coal Transportation in Shanxi Province

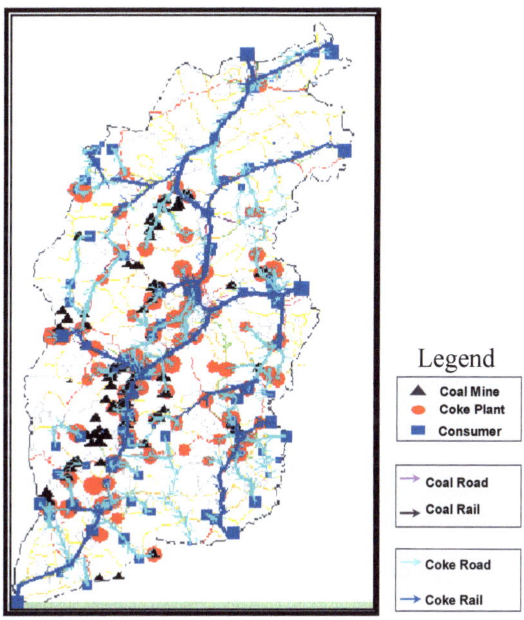

Photo 3: Coke Transportation in Shanxi Province

Photo 4: Coke Chimneys Near Jiexiu, July 2000

Photo 5: Closed Indigenous Coke-Oven Facility, South of Jiexiu, August 1998

Photo 6: Coke Workers Pushing Cover onto Modified-Indigenous Coke Oven, August 1998

TECHNOLOGY: INDIGENOUS AND MODIFIED INDIGENOUS COKE OVENS

Photo 7: Modified Indigenous Coke Ovens, January 12, 2004

Photo 8: Beehive Oven with Green Chimney, July 31, 2002

Photo 9: Sansheng JKH Ovens: Worker Sweeping, July 8, 2000

Photo 10: Qingxu Slot Ovens, with Oven Doors for New Battery in Foreground, July 2001

TECHNOLOGY: MODIFIED INDIGENOUS COKE OVENS AND SLOT OVENS

Photo 11: Coke Push—Antai, July 8, 2000

Photo 12: Cooling Wharf—Antai, July 8, 2000

Photo 13: Quenching Tower–Antai, July 8, 2000

TECHNOLOGY: SLOT OVENS

Photo 14: Worker Cleaning Oven (Sanjia SJ-96), July 31, 2002

Photo 15: Sealing Coke Oven (Sanjia SJ-96) Jiexiu, July 31, 2002

Photo 16: New Sanjia Ovens with Heat Recovery Jiexiu, July 31 2002

TECHNOLOGY: CLEAN COKE OVENS

Photo 17: Houma Coke Ovens,
January 12, 2004

Photo 18: Gangyuan Clean Coke
Ovens, January 10, 2004

Photo 19: Qingxu Yingxian Clean
Coke Ovens, July 31, 2002

Photo 20: Qingxu Yingxian Clean Coke
Ovens with EPB
(Environmental Protection Bureau)
Approved Blue Chimney, July 31, 2002

TECHNOLOGY: CLEAN COKE OVENS

Photo 21: Metallurgical and Foundry Coke, Sanjia, July 9, 2000

Photo 22: Workers Stacking Foundry Coke, Sanjia, July 31 2002

Photo 23: Making Pedestrian Tiles with Waste, July 7, 2001

METALLURGICAL AND FOUNDRY COKE AND WASTE PRODUCTS

Photo 24: Fushun Surface Coal Mine Rail System, August 6, 2002

Photo 25: Dongshan Workers at Mine Mouth, July 6, 2000

Photo 26: Qingxu Truck-Weighing Station, January 10, 2004

Photo 27: Houma Unloading Coal, January 12, 2004

Photo 28: Top-Loading Coal—Clean Coke Oven Qingxu YingXian, July 31, 2002

ENERGY: COAL MINING AND TRANSPORTATION

Photo 29: Coal Washing, July 7, 2001

Photo 30: Workers Testing Coal after Washing, July 7, 2001

ENERGY: COAL PREPARATION

Photo 31: Peasants with Horses Hauling Bricks, Sansheng Coke Plant, July 8, 2000

Photo 32: Small Tractor Used to Haul Coke at Plant, July 8, 2001

Photo 33: Peasants with Tractors Waiting to Weigh Coke, August 1999

TRANSPORTATION

Photo 34: Coke Train Near Jiexiu, July 31, 2002

Photo 35: Foundry Coke Truck Near Jiexiu, January 12, 2004

Photo 36: Workers' Motorcycles Sanjia, Jiexiu, July 31, 2002

Photo 37: Workers Leaving Meijing Coke Plant, January 10, 2004

TRANSPORTATION

Photo 38: Coke Worker Loading 20-tonne Coke Container, Jiexiu, July 9, 2000

Photo 39: Coke Workers Loading 20-tonne Coke Container, Jiexiu, July 31, 2002

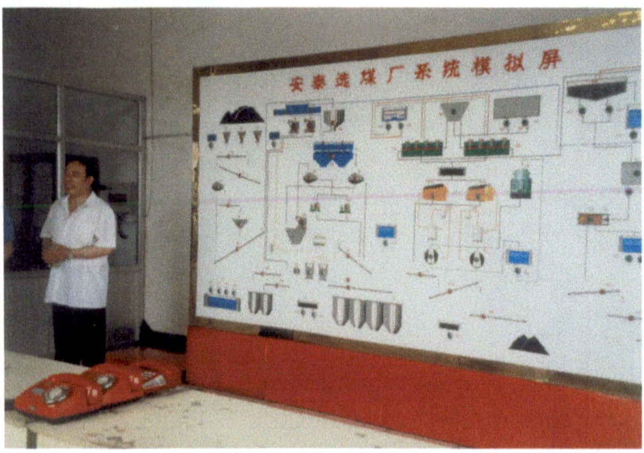

Photo 40: Antai Computer Control Room, July 7, 2001

TRANSPORTATION

Photo 41: Workers Sealing Top of Coke Ovens--Fushun Iron and Steel, August 6, 2002

Photo 42: Safety Sign at Dongsheng Coke Plant, January 11, 2004

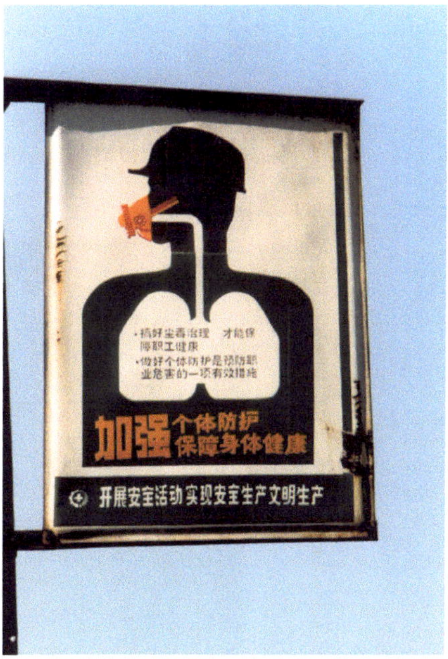

Photo 43: Black Lungs Sign at Taiyuan Coal Gasification Plant, August 1999

ENVIRONMENT

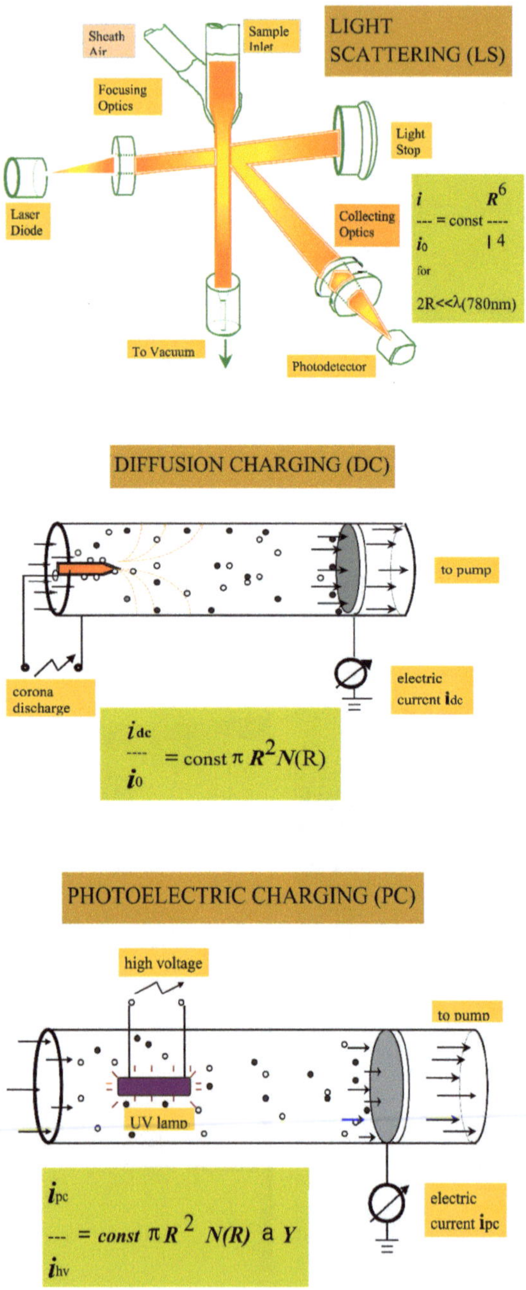

$$\frac{i}{i_0} = \text{const}\,\frac{R^6}{\lambda^4} \quad \text{for } 2R \ll \lambda\,(780\text{nm})$$

$$\frac{i_{dc}}{i_0} = \text{const}\,\pi R^2 N(R)$$

$$\frac{i_{pc}}{i_{hv}} = const\,\pi R^2\, N(R)\, a\, Y$$

Note: LS = Light Scattering; DC = Diffusion Charging; PC = Photoelectric Charging.

Photo 44: Air-Pollution Sensors

ENVIRONMENT

Photo 45: Green Grass and Trees, Antai Cokemaking Plant, July 7, 2001

Photo 46: Qian Zhiqiang Inspecting Three Pollution Sensors, Qingxu, January 10, 2004

Photo 47: Diesel Truck With Cover Hauling Coal, Qingxu, January 10, 2004

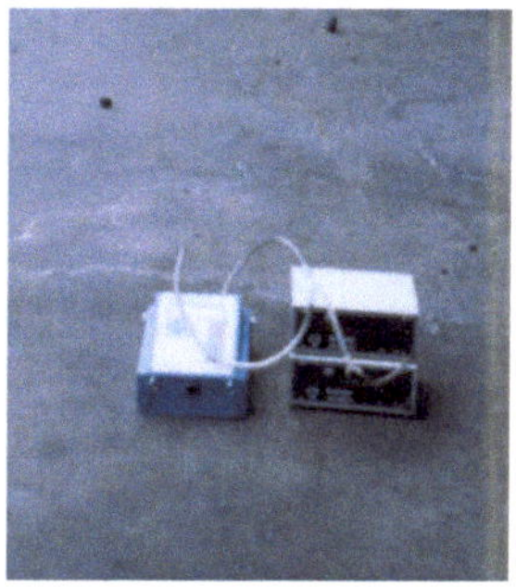

Photo 48: Three Particulate Pollution Sensors, Qingxu, January 10, 2004

ENVIRONMENT

Photo 49: Coke Plant and New Construction, Antai, July 8, 2000

Photo 50: Coke Push with Trees Lining Street by Coke Battery, July 9, 2000

Photo 51: Water Recycling Settlement Basin, July 9, 2000

ENVIRONMENT

APPENDIX B

ACRONYMS

AC	alternating-current
AGS	Alliance for Global Sustainability
B_aP	Benzopyrene-3.4
CAS	Chinese Academy of Science
CEEPR	Center for Environmental and Energy Policy Research
China	People's Republic of China
CIS	Center for International Studies
CNC	condensation nucleus counter
CO	Carbon Monoxide
CO_2	Carbon Dioxide
COG	Coke-Oven Gas
CSY	China Statistical Yearbook
DC	Diffusion charging
DMA	Differential mobility analyzer
DQJ-50	Daogu Qingjie Lu with a maximum capacity of 50 ten thousand tonnes
DUSP	Department of Urban Studies and Planning
EIOT	Extended Input-Output Table
ETHz	Swiss Federal Institutes of Technology, Zürich
GDP	Gross Domestic Product
GFCF	gross fixed capital formation
GIS	Geographical Information System
GPSS	GIS-based Planning Support System
gsce	Grams of standard coal equivalent
GUI	Graphic User Interface
H_2S	Hydrogen Sulfide
HH	Household
HRO	Heat-Recovery Oven
HSPH	Harvard School of Public Health
ISO14000	International Standards Organization 14000
ISO9000	International Standards Organization 9000
ISS	Institute of System Sciences
JKH89-I	JKH-89 ovens designed by Jiexiu Erji Gongsi in 1989, Model I.

Karen R. Polenske (ed.), The Technology-Energy-Environmental-Health (TEEH) Chain in China: A Case Study of Cokemaking, 179–181.

JKH89-II	JKH-89 ovens designed by Jiexiu Erji Gongsi in 1989, Model II.
JKH-97	JKH-89 ovens designed by Jiexiu Erji Gongsi in 1997,
JNK43-98D	Designed by the Anshan Cokemaking & Refractory Engineering Consulting Corporation in 1998, 4.3 meters high, with tamping (Daogu).
JX-1	Jiexiu-1
kg	Kilogram
km	Kilometers
LS	Light scattering
LL Oven	Luliang Oven
MACT	Maximum Achievable Clean Technology
M^3	Cubic meters
MIT	Massachusetts Institute of Technology
NEPA	National Environment Planning Agency
nm	nanometers
PM_{10}	Particulate Matter (up to 10 micrometers in size)
91	1991 type
NO_x	Nitrogen Oxide
NSFC	National Natural Science Foundation of China
ODPHSP	Occupational Disease Prophylactic and Therapeutic Hospital of Shanxi Province
OTA	Office of Technology Assessment
PAH	Polycyclic aromatic hydrocarbon
PC	Photoelectric charging
Plant-Min	2000 Plant-Minimization Scenario
PPAH	Particle-bound Polycyclic aromatic hydrcarbon
PX Oven	Pingxiang Oven
QRD-2000	Qingjie Re Daogu (in Chinese) 2000
QRD-2002	Qingjie Re Daogu (in Chinese) 2002
RMB	Renminbi
SDA	Structural Decomposition Analysis
SPGPSS	Shanxi Province Geographic Planning Support System
SSDA	Spatial Structural Decomposition Analysis
SJ-96	Sanjia 1996 type
SO_2	Sulfur Dioxide
SO_x	Sulfur Oxide
SOE	State-Owned Enterprise
Tce	Tonnes of Coal Equivalent
TEEH Chain	Technology-Energy-Environment-Health Chain
TJ-75	Taiyuan Jixiehua 1975
TJL4350D	Taiyuan Jixiehua Lu, mechanized oven that is 4.3 meters high, 500 mm wide, and features coal tamping
Transport-Min	2000 Transport-Minimization Scenario
TFP	Total factor productivity
TSP	Total Suspended Particulates
TUT	Taiyuan University of Technology
TVE	Township-and-Village Enterprise

TVG	Township and Village Government
2000 Base	2000 Base Scenario
2000 Plant-Min	2000 Plant-Minimization Scenario
2000 Transport-Min	2000 Transport-Minimization Scenario
UNIDO	United Nations Industrial Development Organization
USD	U.S. Dollars
UT	The University of Tokyo
YX-21QJL-1	Yingxian 21 Shiji Qingjie Lu-1, in Chinese; that is, Yingxian 21 Century Clean Oven-type 1, in English

INDEX

D

E

P

Q

R

S

T

U

Alliance for Global Sustainability Series

1. F. Moavenzadeh, K. Hanaki and P. Baccini (eds.): *Future Cities: Dynamics and Sustainability*. 2002 ISBN 1-4020-0540-7
2. L. Molina (ed.): *Air Quality in the Mexico Megacity: An Integrated Assessment*. 2002 ISBN 1-4020-0452-4
3. W. Wimmer and R. Züst: *ECODESIGN Pilot. Product-Investigation-, Learning- and Optimization-Tool for Sustainable Product Development with CD-ROM*. 2003 ISBN 1-4020-0965-8
4. B. Eliasson and Y. Lee (eds.): *Integrated Assessment of Sustainable Energy Systems in China. The China Technology Program. A Framework for Decision Support in the Electric Sector of Shandong Province.* 2003 ISBN 1-4020-1198-9
5. M. Keiner, C. Zegras, W.A. Schmid and D. Salmerón (eds.): *From Understanding to Action. Sustainable Urban Development in Medium-Sized Cities in Africa and Latin America*. 2004 ISBN 1-4020-2879-2
6. W. Wimmer, R. Züst and K.M. Lee: *ECODESIGN Implementation. A Systematic Guidance on Integrating Environmental Considerations into Product Development*. 2004 ISBN 1-4020-3070-3
7. D.L. Goldblatt: *Sustainable Energy Consumption and Society*. Personal, Technological, or Social Change? 2005 ISBN 1-4020-3086-X
8. K.R. Polenske (ed.): *The Technology-Energy-Environment-Health (TEEH) Chain in China*. A Case Study of Cokemaking. 2006 ISBN 1-4020-3433-4

MIX
Papier aus verantwortungsvollen Quellen
Paper from responsible sources
FSC® C105338

If you have any concerns about our products,
you can contact us on
ProductSafety@springernature.com

In case Publisher is established outside the EU,
the EU authorized representative is:
Springer Nature Customer Service Center GmbH
Europaplatz 3, 69115 Heidelberg, Germany

Printed by Libri Plureos GmbH
in Hamburg, Germany